SpringerBriefs in Applied Sciences and Technology

PoliMI SpringerBriefs

Springer, in cooperation with Politecnico di Milano, publishes the PoliMI SpringerBriefs, concise summaries of cutting-edge research and practical applications across a wide spectrum of fields. Featuring compact volumes of 50 to 125 (150 as a maximum) pages, the series covers a range of contents from professional to academic in the following research areas carried out at Politecnico:

- Aerospace Engineering
- Bioengineering
- Electrical Engineering
- Energy and Nuclear Science and Technology
- Environmental and Infrastructure Engineering
- Industrial Chemistry and Chemical Engineering
- Information Technology
- Management, Economics and Industrial Engineering
- Materials Engineering
- Mathematical Models and Methods in Engineering
- Mechanical Engineering
- Structural Seismic and Geotechnical Engineering
- Built Environment and Construction Engineering
- Physics
- Design and Technologies
- Urban Planning, Design, and Policy

Anna Faiella

Learning from Disasters to Enhance Resilience of Urban Complex Systems

Stress Testing Cities and Urban Plans for Resilience

POLITECNICO
MILANO 1863

Springer

Anna Faiella
Department of ABC
Politecnico di Milano
Milan, Italy

ISSN 2191-530X ISSN 2191-5318 (electronic)
SpringerBriefs in Applied Sciences and Technology
ISSN 2282-2577 ISSN 2282-2585 (electronic)
PoliMI SpringerBriefs
ISBN 978-3-032-07355-6 ISBN 978-3-032-07356-3 (eBook)
https://doi.org/10.1007/978-3-032-07356-3

This Springer imprint is published by the registered company Springer Nature Switzerland AG
The registered company address is: Gewerbestrasse 11, 6330 Cham, Switzerland

Foreword

The book *Learning from Disasters to Enhance Resilience of Urban Complex Systems: Stress Testing Cities and Urban Plans for Resilience* is the result of extensive and tenacious research on how to address recognized gaps in the journey towards safer and more resilient cities.

As mentioned by the author, it combines three streams of research: spatial planning as a non-structural measure to prevent and mitigate disaster risk and adapt to climate change impacts, learning from post-disaster damage and loss data, and complementing risk and resilience management.

Blending these three streams required a detailed exploration of each topic and a review of available literature. Most importantly, it required making explicit the interconnections between the three streams and explaining how they could be effectively used to enhance the understanding of both vulnerabilities and resilience of cities, providing practical guidance to planners.

Spatial planning is a key component of strategies necessary to reduce the level and extent of damage resulting from natural extremes, included those that are exacerbated or created by climate change.

The discourse on climate change adaptation and resilience has increased the interest of researchers and public administrations in improving the urban fabric and built environment in the face of natural and man-made stressors. As explained in the book, resilience provides a comprehensive conceptual umbrella under which both pre-event and post-event provisions can be hosted, with the long-term view that characterizes urban and spatial planning as a practice. Traditional risk assessment is still key to identifying and mapping hazards, exposure, and vulnerabilities and designing mitigation measures that can reduce expected damage to people, assets, and infrastructure. Resilience assessment and management are needed to prepare for the unexpected, balancing vulnerabilities that are hard to address, and improving response and recovery. Resilience requires that rebuilding and rehabilitation capitalize on lessons learned from past damage to reduce vulnerabilities to future extremes.

Spatial planners need guidance for locational and zoning decisions that involve higher population densities and functions, aiming to minimize exposure to hazards and risks. In the search for operationalization, two tools have captured the author's attention, who decided to test them on the case study of Benidorm. The idea of stress testing infrastructures and urban planning for resilience has been debated with Igor Linkov[1] who proposed it for the first time to our knowledge. In Benidorm, it proved effective in highlighting the challenges faced by past and current city planning and management in a multi-hazard environment. Benidorm is paradigmatic of unsafe urbanization trajectories that have characterized the coastline of some European countries. It was selected thanks to the collaboration with Jorge Olcina Cantos,[2] who hosted the Author as a visiting researcher. Extreme events that hit Spain, Italy, Greece, and Central Europe in the past few years unveiled the different layers of vulnerabilities that characterize our cities and settlements despite good intentions, investment in sustainability, and careful legislation on urban and spatial planning.

One of the stress tests conducted on Benidorm relates to the availability of data to support complementary risk and resilience assessment and management. The results show that while good quality data is available in some domains, such as insurance compensation for past damage, data is lacking in other aspects, such as coherent multi-hazard mapping and vulnerability assessments for potential extremes.

The underlying idea of this stress test is that evidence-based planning is crucial for designing and redesigning new and old city areas, requiring good quality data tailored to the local context. This includes geospatial, demographic, economic, lifeline, and transport data combined with information on hazards and risks affecting urban systems and communities. The trial conducted in the Risk Data Hub platform developed by the Disaster Risk Management Knowledge Centre of the European Commission, populating it with data from the Benidorm case, has been important to exploit the system's analytical potential and to prove that some level of standardization is possible for other cities to explore.

We are aware of the limitations and critiques of data-driven urban and spatial planning and the risk of technocratic drift in current practices and research on sustainability[3] and resilience, especially with recent advancements in AI. Such awareness is important to avoid overlooking the fundamental social, economic, political, and governance constraints that have hampered safer urban development and planning for resilience. Still, the data presented and used to develop better knowledge on risk components and factors that may increase or decrease the resilience of cities and settlements is crucial for supporting also enhanced participatory practices. These practices involve citizens and community stakeholders bringing in their knowledge and experience of the local environment and past events. Sharing data and knowledge between communities, stakeholders, and experts in different disciplines can produce more nuanced and context-sensitive evaluations of hazards, exposure, and vulnerabilities and explore novel solutions for more resilient settlements.

---

[1] See Linkov, I., Trump, B. D., Trump, J., Pescaroli, G., Hynes, W., Mavrodieva, A., & Panda, A. (2022). Resilience stress testing for critical infrastructure. *International Journal of Disaster Risk Reduction, 82.*

[2] Sánchez-Almodóvar, E., Olcina-Cantos, J., Martí-Talavera, J., Prieto-Cerdán, A., & Padilla-Blanco, A. (2023). Floods and adaptation to climate change in tourist areas: management experiences on the coast of the province of Alicante (Spain). *Water*, *15*(4), 807.

[3] Joss, S. (2017). *Sustainable cities: Governing for urban innovation.* Bloomsbury Publishing.

At last, one final note on the idea of innovation in urban and spatial planning that is an important underpinning of Anna Faiella's book. The discourse on sustainability and innovation, both technological and data-driven, has been explored in literature, highlighting both positive and problematic aspects. However, innovation is essential to make communities less vulnerable and cities more resilient to extreme events and climate change impacts. This requires better connections between researchers in different disciplines and government sectors to leverage each other's investments with longer-term time horizons in mind. Therefore, innovation is needed not only in the more competent use of information systems and datasets, but also in the way knowledge is formed, maintained, and shared among different stakeholders and citizens. This is a fundamental premise for establishing a foundation for more cross-sectoral governance, which will lead to changes in unsustainable urbanization practices and ultimately result in less damage and trauma for people affected by extreme events in the future.

Prof. Scira Menoni
Department of ABC
Politecnico di Milano
Milan, Italy

Acknowledgments My profound gratitude goes to my mentor Prof. Scira Menoni, not only for her continuous and invaluable support during my career, but especially for always believing in me. I am extremely grateful to her for guiding me since my post-graduate studies, providing me with significant opportunities to grow and deepen my knowledge in the field of disaster risk reduction.

I would like to thank Prof. Jorge Olcina-Cantos for hosting me as a visiting researcher at the University of Alicante and Dr. Christina Corbane, as well as all the team of the Risk Data Hub of the Joint Research Center of the European Commission. I am extremely grateful for their support in the development of my research.

Moreover, I would like to express my gratitude to: Dr. Igor Linkov, Dr. Grazia Concilio, Dr. Christer Pursiainen and Dr. Galvin Smith, for their insightful review of the Ph.D. thesis that has been the base for this book.

Competing Interests The author has no competing interests to declare that are relevant to the content of this manuscript.

Contents

Chapter 1
Spatial Planning, Disaster Risk Reduction and Resilience Management: A Framework

Abstract This first chapter discusses how a better relationship between disaster risk reduction and spatial planning could represent an opportunity for preventing and mitigating disaster impacts, proposing the integration of resilience management being an approach able to deal with the complexities of contemporary systems. Following the introduction of resilience intended as a as process which learns from past experiences, the relevance of post-event damage and loss data potential and their different uses in the field of disaster risk management are addressed in the chapter.

1.1 The Relationship Between Spatial Planning and Disaster Risk Reduction

The discipline of spatial planning is here intended as complex and multidisciplinary process aimed at addressing strategically future spatial distribution of societal activities in a given geographic area by creating a rational territorial organization and spatial arrangement of land uses to achieve social and economic objectives while safeguarding the environment (Wegener 1998; Yoshida et al. 2020). Spatial Planning involves different forms of planning and approaches such as urban planning and land-use planning; and, makes use of information and knowledge developed from a vast array of fields, from architecture and engineering to demography, law and many others, thus reflecting the complexity of both "the objects of study and practice" (Huxley and Inch 2020). As written in Nowak et al. (2023), considering the numerous terms linked to spatial planning used and intended in different ways in every country, spatial planning can be understood, in other words, as a process aimed at regulating permitted activities in specific locations (Van den Hoek 2020). Due to its nature, the purpose of planning is strictly linked with social processes and policy (Wegener 1998) and therefore constantly confronted with changing needs, to which spatial planning should respond flexibly. Amongst the many purposes for which spatial planning is performed, risk reduction and prevention for both existent and future anthropized environment should be considered as key. Practices of spatial planning shape exposure and vulnerability to disaster, making disaster

A. Faiella, *Learning from Disasters to Enhance Resilience of Urban Complex Systems*, PoliMI SpringerBriefs, https://doi.org/10.1007/978-3-032-07356-3_1

risk reduction principles fundamental to be mainstreamed in spatial planning. Risk is here intended as a function of hazard, exposure and vulnerability. Hazard is a process or phenomena which could cause damage and losses and it can have natural, anthropogenic or socio-natural origins; its combination with exposure, such as the number of people or types of assets located in hazard-prone areas, and vulnerability intended as the propensity or the susceptibility of an individual, a community, assets or systems to the impacts of hazards generate the final risk (UNDRR 2017). Disaster risk reduction is the policy objective of disaster risk management, and it is "aimed at preventing new and reducing existing disaster risk and managing residual risk, all of which contribute to strengthening resilience and therefore to the achievement of sustainable development" (UNDRR 2017).

Natural events, are dynamic processes of the Earth, and they become disasters when they adversely impact communities leading to human, material, economic and/or environmental damage and losses. The significance of the impacts is strongly related to exposure and vulnerability of the context in which they occur. Both those components of risk are the result of actions and measures related to spatial planning and land use management. Often the development of the built environment and its associated planning process become cause of disasters when they increase exposure and vulnerabilities to existing hazards or they generate new ones (Esteban et al. 2011; Galderisi and Profice 2012; Wamsler 2008). Spatial planning is therefore a fundamental instrument for risk reduction and prevention for both existent and future anthropized environment. Spatial planning fall within the strategies and actions aimed at lessening or preventing, when possible, the adverse impacts of natural hazards. Those strategies and actions known as non-structural mitigation measures—differently from the structural ones which mainly consist in engineering artefacts—aim at addressing, through policies and laws, the improvement of social, economic and organizational settings to reduce disaster risks and impacts enhancing the resilience of communities and assets in the face of hazards (Coppola 2006; Dynes 1993; Rossi et al. 2013). Disaster risk management considered as a practice to apply across different societal sectors (UNISDR 2015), can be mainstreamed in spatial planning mutating it in a proactive mean to reduce and prevent risks.

For a sustainable future, communities and their representatives must "make informed choices" through plans and programs aimed at exploiting areas that will not endanger the long-term functioning of their systems (Burby 1998). However, too often, areas prone to natural hazards are characterized by certain features which make them appealing for development. Numerous cities have developed in the proximity of water bodies, rivers or seas which constitute poles of economic developments for transportation of goods and people, expansion of tourism and other activities such as fishing. The Spanish city of Benidrom in the Alicante Province provided here as a case study is a clear example of such dynamics (see Chap. 3). Structural mitigation measures, such as dams, flood storage reservoirs, levees, sea walls are usually implemented to reduce hazard impacts in those flood prone areas. However, such strategies are costly and do not provide total protection; instead, they might create a false sense of safety encouraging development in hazardous areas and ultimately increasing the likelihood of losses occurrence due to higher exposure. In this context proper land use can reduce potential damage and losses by constraining development in hazardous areas (limiting excessive developments and/or addressing specific design approach of hazard-resistant constructions in hazardous prone areas) while preserving environmental settings. A suitably conducted planning process foresees and allows a combination of those land use approaches. To be effective spatial planning should be tailored for the context in which it is applied and used in conjunction with other strategies such as emergency planning, early warning and economic tools as incentives and insurances (Burby 1998). In the same book, in chapter two Rutherford H. Platt (1998, p. 30) writes "The beginning of wisdom concerning disasters is the recognition that choices must be made, both in the original layout of new settlements in hazardous areas, and in recovery after a disaster strikes. Indeed, the aftermath of a disaster offers a critical opportunity to review the choices available and possibly to select modification of the status quo ante to achieve greater protection from future disasters." a concept still strongly valid in current times considering that even if knowledge and practices of disaster risk management have improved in the past decades, natural phenomena still cause important damage and losses in many countries, as demonstrated by the recent devastating 7.8 magnitude earthquake and nearly as strong aftershock occurred on 6th February 2023 near the Turkey-Syria, or the massive floods occurred in Libya in September 2023 due to Storm Daniel, as well as the October 2024 Spanish floods which affected the Valencian area. Disaster risk management needs to be linked more consistently to urban and land use planning (Okada 2004), which has a key role in addressing transformations that can increase resilience through adaptation and mitigation measures to limit the impact of disasters.

In the Fifties, the most common approach to dealing with risks was mainly focused on the solely hazard component defining risk as a function of probability and expected damages (White 1945; Oliver-Smit 1996; Adams 1995). Later on, such approach, which considers only the natural phenomena as the responsible for the magnitude of the impacts, started to change including other factors related to the context exposed to the hazard in risk analysis. Along with enhanced risk assessments as function of hazard, exposure and vulnerability also risk reduction strategies have significantly evolved in the past 40 years. Science and practices witnessed a shift from reactive schemes (relief and response) to more proactive ones (preparedness and prevention) which consider risk management and mitigation placing emphasis to the pre-disaster planning (UNDRO 1979) in order to avoid or reduce the potential damage occurrence in case of hazardous events. Such change led to growing awareness by governments that pre-disaster planning should be an integral part of national development policy and that an enhanced formulation and enforcement of land-use policies and plans and appropriate building codes, are key factors for the development of safer communities (UNDRO 1979). Spatial planning is key for the achievement of disaster risk reduction, having the potential to influence physical, spatial and functional collective outcomes that can be by themselves a mean to reduce the impact of disaster and enhance resilience (March and Kornakova 2017). The integration of spatial planning into disaster risk reduction has become increasingly solicited by several policies (UN HABITAT 2016, UNDRR 2017; European Green Deal COM/2019/640, The Territorial Agenda 2030; EU Strategy of Adaptation to Climate Change COM/2021/82;

European Flood Directive Directive 2007/60/EC etc.) and governments, as the awareness of the connection between the characteristics of urban spaces, communities organizations and risk has been constantly evolving (March and Leon 2013). In the last years, several academic works have been dealing with disaster risk reduction and resilience mainstreamed into urban planning practices thanks to various international initiatives (starting from the Hyogo Framework for Action 2005–2015, followed by the Sendai Framework for Disaster Risk Reduction 2015–2030 in which priority 3 calls directly for "investing in disaster risk reduction for resilience"; and other examples as the 2030 Agenda for Sustainable Development; the Paris Agreement; and at the European level the Climate Law EU 2021/1119; the European Green Deal; and the EU Strategy on Climate Change Adaptation) that have highlighted the importance of aligning the latter to reach the common objectives that they intrinsically share. Though, communication between specialists and professionals focused on different disciplines and with different backgrounds, in addition to the complexity of practically shifting from the traditional reactive policies to proactive ones, interfere with the implementation of scientific achievements in the field of natural hazards in spatial plans (Galderisi and Menoni 2007). Some European projects aimed at proposing tools and good practices to enhance the integration of disaster risk reduction and spatial planning (ARMONIA,[1] 2004–2007; RAMSES,[2] 2012–2017; ENSURE, 2008–2011 funded under the EU VII FP) and various scholars addressed the importance of land use planning as primary mean for risk prevention and reduction; the development of the built environment and its associated planning process become cause of disasters increasing exposure and vulnerabilities to existing hazards or even generating new ones if not properly integrated with prevention (Esteban et al. 2011; Galderisi and Profice 2012; Wamsler 2008).

Spatial planning practices may amplify the consequences of natural disasters if natural dynamics are not taken into account. A better relationship between disaster risk reduction, sustainable development and planning policies could represent an opportunity for preventing and mitigating disaster impacts. Even though spatial planning and disaster risk management are characterized by a socio-spatial synergy since they deal with the same objects and both are based on knowledge of the territory, they represent two distinct processes, each administrated by specific laws/frameworks and distinct institutional organizations (Kammerbauer 2013; Mejri 2019). First, as mentioned by Kammerbauer urbanistic discipline focuses on spatial, social, and institutional aspects of urban everyday life, while disaster risk reduction approaches have to deal with disruptions of normality with strategic interventions to implement before, during and after a disaster occurs. Second, the two disciplines produce different tools and instruments such as plans for spatial planning, and programs and actions for disaster risk management. Spatial planning focuses on future developments over different spatial and temporal scales for both built and natural environment and settled communities, producing a set of visions and instruments to guide those transformations following normative and policies delineated at country and

[1] http://www.t-6.it/project/armonia/

[2] https://ramses-cities.eu/home/

regional levels (Secchi 2011) which are independent of any catastrophic event that could occur. Besides, spatial planning is part of governance structures, while the disaster risk reduction management and responsibilities are shared across different organizations and authorities making its administration unclear and more complex. Despite this context, disaster risk reduction knowledge can become a base element to generate spatial planning design solutions aimed at reaching the objective of reducing or avoiding the creation of risks with a spatial focus on physical components while integrating social, economic and environmental considerations. Nonetheless, planning is a form of governance (March and Kornakova 2017) which focuses on collective benefits creating advantageous spatial arrangements of both physical and functional systems of the area under analysis improving social, economic and ecological outcomes through physical processes.

Therefore, disaster risk reduction and planning both aim at balancing and managing current circumstances with future goals and possibilities (Hopkins 2001) to enhance societal settings. Spatial planning, being a tool that should produce and use knowledge of territories, is one of the practices of paramount importance at the core of disaster risk management and reduction. Government authorities should consider spatial planning as a pre-disaster prevention, mitigation, and preparedness process which in all its phases of intervention aims at reducing risks with land policy formulation that applies tailored legal and technical regulations (Roy and Ferland 2015). Effective outcomes cannot be met by one discipline alone; social, institutional and spatial features of the urbanized environments, vulnerabilities and hazards should be handled with an interdisciplinary and multi-temporal approach, which considers pre-existing and post-event conditions. A more comprehensive and operational interchange between the two disciplines in order to appreciate the results of the strategies of both is necessary. However, the integration of disaster risk reduction strategies in spatial planning is challenging in practice. The lack of integration between territorial governance processes and risk mitigation is one of the elements limiting the effectiveness of a preventive spatial planning (Wamsler 2006), in addition to the complexity related to the interaction and communication processes within different disciplinary domains (Galderisi and Menoni 2007; Gazzola 2019). Ideally, planning should identify problems and/or opportunities and provide alternative goals and objectives, which effects, once implemented, should be monitored to revise, re-adjust and re-direct the process (or its goals) in order to meet its actual goal and obtain enhanced outcomes. But practices of spatial planning are often regulated by instruments that do not allow easily to reconsider and improve such process due to their highly procedural nature. Spatial planning is regulated by static and complex instruments and tools, making hard the implementation and integration of strategies which are aligned with the dynamic needs of a system; its practices are complex and the operational settings are not generally oriented to disaster risk reduction. Both spatial planning and disaster risk reduction seek enhanced future outcomes, even though, sometimes, they act on different time rages (for instance the process of urban development unfolds over decades, while disaster risk reduction might focus on obtaining results on shorter terms), they are equally aimed at addressing an enhanced collective future based on the understanding of the past, present and plausible future. As highlighted by March and Leon (2013) and Galderisi and Menoni (2007), despite commitment of some planners, innovative concepts and new policies aimed at the integration between planning and disaster risk reduction, it has been difficult to translate it into practice and few guiding principles have been delineated to support such practice. In many cases, the required integration remains only formally implemented, leaving the two processes separated de facto. Greater convergence has been attempted between planning and resilience, understood as a process that should be simultaneously developed across multiple overlapping dimensions such as the built environment, social dynamics, governance (Menoni 2020).

Considering spatial planning as proactive means to reduce risks and enhance resilience, a variety of tools are available to address and manage more cautious developments. Examples include taxation and fiscal policies, development regulations which govern the location, type, and intensity of new development such as zoning which limits the development in hazardous areas and regulations that require adequate mitigation measures in new urban development as building codes, drainage systems, relocation of existing development at risk, proper location of infrastructures, design standards, and public information programs that raise risk awareness. Through a proper combination of those development tools, governments can enhance prospects for a sustainable future. Proper planning can alter the overall damage function through time by modifying the development processes in hazardous areas (Dawson et al. 2011). Planning approaches aimed at mitigating risk are mainly subdivided in location and design approach and make use of different operational tools to concretely address risk reduction. The location approach limits urban development, acting on location and land use choices while preserving environmental values; while the design approach provides structural interventions to guarantee the reduction of physical vulnerability, the control of exposure and the increase of urban resilience (Burby 1998). Following the initial work of Bolton et al. (1986), Menoni (2006, 2012) identifies and classifies existing planning tools for risk prevention used to satisfy the aforementioned planning approaches. Such tools can be subdivided in four main groups as follow:

- Zoning: creation of safety distances or separation buffers between vulnerable systems and the hazard. The implementation of this tool is possible in the case of new expansion or urban redevelopment projects, but it is not very effective in already consolidated urban areas;
- Areal relocation: suitable tool for intervening in already consolidated areas, this involves the transfer of specific urban functions from high risk areas to others characterized by lower risk (Perry and Lindell 1997; Menoni and Pesaro 2008);
- Technical and structural interventions: envisage the implementation of specific engineering interventions to reduce physical vulnerability;
- Non-regulatory tools: used to achieve prevention indirectly by addressing preventive and precautionary behaviours influencing/disincentivising possible negative behaviours mainly in the real estate and insurance market (insurance, taxation, incentives, mostly known as economic tools).

Those tools, appropriately combined, guarantee the achievement of disaster risk reduction. Table 1.1 depicts a more detailed subdivision and consideration of the different sub-typologies of such tools.

However, the current interplay between DRM and spatial planning is not fit for the purpose of managing the challenges faced by complex urban systems in multi-risk environments with rising exposure and vulnerabilities. Risk and resilience management seem to be both key elements to be suitably integrated in order to embrace existing challenges related to disaster risk reduction and planning and overcome some of the existing gaps in the field of risk management (Menoni et al. 2024). The

Table 1.1 Spatial planning tools for disaster risk reduction achievement

Zoning	Subdivision standards	Building codes
Special seismic study zone	Performance standards for sensitive lands	Supplemental seismic standards
Open space/conservation zones	Standards and regulations for new development areas	Standards to be introduced in building codes
No-buildings zones	Regulations and codes for urban renewal	Seismic standards for retrofitting residential buildings
Hazardous building relocation	**Strategic public facilities**	**Lifelines**
Abatement ordinance for risky plants	Abatemet ordinance for public facilities in dangerous areas	Substitution of old lifelines in dangerous areas
Increase the mutual distance between dangerous facilities and residential areas	Relocation of strategic facilities from dangerous areas	Retrofitting of lifelines in the most vulnerable situations
Incentives to relocate industrial plants in dangerous sites	New public facilities in safer areas	New infrastructures in safer areas
Real estate disclosure	**Property/development rights**	**Taxes, incentives**
Obliging contractors discole risk to potential buyers	Property voluntary acquisition or expropriation	Tax benefits for those who retrofit their house
Particular conditions on selling and buying contracts	Purchase or expropriation of developments rights	Incentives for those relocating from dangerous areas
Insurance policies	**Environmental impact studies**	**Mapping, research, studies**
Insurance programs for goods exposed to risks	Including risk analysis in environmental impact assessment studies	Mapping programs in risky areas
Insurance programs not only for private citizens but also for local/provincial authorities		Surveys and research programs in risky areas
		Creation and updating of historic data on past disasters

Source Menoni (2006)

following section discusses opportunities that arise from the integration of risk and resilience management.

1.2 Integrating Resilience Management and Disaster Risk Reduction

The concept of resilience has become increasingly important in disaster risk reduction programmes and also linked to cities and urban planning (Mohammadi et al. 2023; Mansavi et al. 2019; Béné et al. 2018). Resilience is a multifaceted concept which has been interpreted in a variety of forms and within several disciplines (Norris et al. 2008; Alexander 2013). Literature offers a vast set of definitions for such a complex concept (Norris et al. 2008; Manyena 2006; MacAskill and Guthrie 2014) and there is no broadly consensus about a unique definition (Klein et al. 2003; Manyena 2006). Recently, the concept has been increasingly used in scientific and political fields related to sustainable development and disaster risk reduction (Sharifi and Yamagata 2018). Between many, UNDRR (2017) defines resilience as "the ability of a system, community or society exposed to hazards to resist, absorb, accommodate, adapt to, transform and recover from the effects of a hazard in a timely and efficient manner, including through the preservation and restoration of its essential basic structures and functions through risk management". Resilience entails the notion of dynamics and complexities of a system and focuses on its capacity to respond and adapt to a stressor or disturbance intended as adverse circumstances that threaten the functioning of the system. Calls to mainstream, improve and boost resilience, a prominent concept nowadays, are made at different levels and for different domains, and efforts are done to operationalize it within policies and, more specifically, in the field of urban resilience (Patel and Nosal 2016).

There is a strong interdependence among the discipline of spatial planning and risk conditions (Gazzola 2019), as well as the concept of resilience, an essential component which is required to be implemented by many policies defined at different governmental levels (Sendai Framework for Disaster Risk Reduction also the Sustainable Development Goals; The 2030 Agenda for Sustainable Development; European Climate Law EU 2021/1119; European Green Deal COM/2019/640). Natural hazards can impact both the physical configuration of a territory and its functional organization, damaging not only physical assets and people but also disrupting socio-economic functions, leading to the opposite outcomes sought by spatial planning which aims at addressing the organization and the development of societal activities. Preventive and proactive planning processes and choices may contribute to reduce risk levels by properly acting on the variables of risk, namely: hazard, exposure and vulnerability. Correct planning practices are also relevant to enhance resilience. Resilient communities are those where their development has been based on sustainable development goals (Godschalk et al. 1998), where disaster risk reduction is a key determinant (UNISDR 2012). In his book "Cooperating with Nature.

Confronting Natural Hazards with Land-Use Planning for Sustainable Communities" Burby (1998, p. 1) writes "by planning for and managing land use to enhance sustainability, we can reduce our vulnerability to disasters". Natural hazards are a major threat to communities' resilience, hampering sustainable development, economy, and social settings. The speed, severity, and complexity of known and unknown changes in climate, the often uneven impact across different areas, and the increasing interconnectedness of societies and their functions, increase the complexity of the approaches, methodologies and practices that aim at managing and reducing risks. Thus, resilience becomes a guiding principle to address those complexities related to climate change adaptation and disaster risk reduction systemically and comprehensively handled, linking the requirement of learning from the past to address the times to come.

Societal sectors functioning relies on products and services provided by highly interconnected systems. Most services and production activities heavily rely on power and telecommunication infrastructures between which, to make an example: the operativity of banks, hospitals, essential administrative services that rely on the operationality of telecommunication networks and on data centers which are prone to failure as a consequence of threats besides cyber, such as natural hazards and increase of temperature due to climate change. Industries and hospitals functioning heavily depends not only on power and telecommunication, but also on transportation infrastructures which if potentially damaged could cause logistics problems such as the impossibility to reach the interested site for workers, essential materials, and rescuers.

In addition to the complexity of our society settings, threats by themselves are difficult to deal with. Territories may be exposed not only to one single hazard but to different ones at the same time. Coexisting hazards are becoming more recurrent or perhaps simply better recognized (Boni et al. 2025). Climate Change is the emblem of hazards interaction, moreover the complexity of our society and the spaces we live in and/or transform according to our needs make enchained events more frequent.

Climate change impacts are altering and intensifying existing risk patterns as proven by changes in temperature and precipitation, more frequent and extreme storms, wildfires, floods, rising sea levels and droughts (IPCC 2018), the degradation of ecosystems increases the vulnerability of populations and physical assets to disasters; meanwhile the number of weather-related disasters per year have increased such as the related damage and losses given the growing concentration of people and economic assets in hazard-prone areas (OECD 2020). A coherent approach towards the current situation contemplates the integration of disaster risk reduction and climate change adaptation in sustainable development (Dazé et al. 2018). However, its success is bounded to good practices in spatial planning at different spatial scales (King et al. 2016). Several are the barriers to the actual integration of climate change perspective within spatial planning processes, as the different spatial scales of action, the lack of a mechanism to coordinate such endeavour as well as the institutional inadequacies related to laws and regulations (Nowak et al. 2023).

The integration into urban policy and practices is possible when principles of resilience are part of citywide development plans; the capacity to manage spatial planning to address a sustainable development while protecting people and their assets, included ecosystems, finds solid ground where the concept of resilience is well anchored. Climate change increases the need for stronger resilience which sees disaster risk reduction developed and implemented through adaptation.

"We live in a complex world" (Wlaker and Salt 2012), dealing with risk in a complex world is a challenge, and this calls for "resilience-thinking" (Chandler 2014). Interaction and interconnection are crucial elements of complexity so that analysing only the components of a system without analysing the relationships between them (and the role of interconnected systems) is not such a suitable approach. Linear reasoning is insufficient to comprise complexity which is not simply reducible to a sum of the parts of the system and where adaptation becomes the solution (Chandler 2014). New strategic and systemic approaches are hence imperative to manage current and future challenges. Adaptive tools built on solid scientific basis should be developed. Thus, a "resilience-thinking" approach, becomes a solution to the problem of governing complexity as written by Chandler (2014) in the title of his book. Yet, only few studies address resilience of urban systems handling them as a whole and taking into account the dependencies and interactions between the different urban components (Rus et al. 2018).

Developing urban resilience is the subject of a global effort and a top priority for international agendas which recognize the importance to create inclusive, safe, resilient and sustainable human settlements. The development and boosting of resilience within urban growth implicitly identifies disaster risk reduction as a practice fundamental also for sustainable development. Most of the global population lives in urban centers, only in Europe 75% of the population lives in urban areas,[3] and this value is expected to grow in the next decades. Numerous cities are located in hazard-prone areas where the severity of the impacts is expected to grow due to climate change UNISDR (2012). Making cities resilient becomes a prerequisite for the current and future times where known and un-known risks hamper the wellbeing of people and the functioning of economic, cultural and environmental systems on a global level. Efforts by both national and local governments towards enhancement of disaster risk reduction processes to build resilience fail to keep up with the rapid urbanization experienced (UNISDR 2012), alongside the limit posed by some realities in which institutional vulnerability slows down the achievement of safer and more resilient urban spaces.

Therefore, nowadays, the focus on resilience seems to be more pertinent than ever. Resilience is fundamental in the present context where complex dynamics take place, as communities and territories are highly interconnected leading to pathways where local disruptions can have widespread impacts which can be experienced even at global level. We have experienced, in the past years, more than ever, the significance of operational continuity of our complex interconnected systems. Adaptive practices and tools able to face nowadays challenges, as also indicated by the Strategic Crisis Management in the Scientific Opinion No.13 of 2022 (GCSA 2022) need to be put in

[3] http://www.statista.com/statistics/270860/urbanization-by-continent/

place so that resilience can be addressed and enhanced, to avoid escalation of negative effects on communities and their territories. Fuchs and Thaler (2018) stress the need to move beyond hazard and risk management to fully integrate resilience and adaptation approaches. The focus of disasters studies was originally more focused on hazard and on the exposed physical environment. However, this approach fails to fully explicate the increasing damage and losses that are increasingly recorded. Consequently, the approach has evolved including consideration of the human facet of disasters focusing on vulnerability and resilience. Such shift is often more in theory rather than in practice. The traditional risk analysis which considers hazard, exposure and vulnerability involves the quantification of each component. As systems become increasingly interdependent, the effects of climate change are still not properly understood, ripple and cascading effects are widespread, and such an approach doesn't seem apt to address current challenges. Often aspects related to human behaviours are not properly understood and recorded, applicable vulnerability parameters are very limited, perception of risk or economic features are hardly considered, damage and losses assessments are based on very aggregated data. More adaptive methodologies and approaches need to be addressed to be able to deal with the current situation. In such a context resilience becomes the "new way forward" as stated by Fox-Lent and Linkov (2018). In a framework where quantitative risk assessment dominates over resilience understanding (Park et al., 2013), resilience concepts should be better integrated with risk management approaches (Linkov et al. 2013).

The present research considers the necessity to shift from risk-based approaches (which aim at identifying components of failing systems) to resilience evaluation, to understand not only the impact of the disruption but also the system temporal capacity to recover and adapt. A more systemic approach allows to focus on different phases subsequent to an event, therefore not only on how a system reacts in the aftermath of the stress, but also on its coping and adaptation capacities. Resilience management in broader terms is a process aimed at enhancing awareness and increasing operational capacity by highlighting the critical components necessary for operations and functioning of the system assessing the criticalities that can hamper operation during the disruption, response and recovery phase (McManus et al. 2007). Considering the features of nowadays societies and their dependence on networks, understanding the capacity of systems to perform throughout disruptive events becomes essential. It is in the complexities of crisis experienced and expected that resilience management manifests its added value, bringing together different disciplines, from disaster management, to planning, social science, economy, policy development and so on, necessary deal with multi-hazards, multi-risk environments and multiple facets of contemporary societies and economies. Even more so as cities management has to guarantee the development of sustainable and resilient cities (Fuchs and Thaler 2018).

1.2.1 Conceptual Definition of a Resilient System

Whilst the analysis of the large body of literature aimed at defining and understanding resilience goes beyond the scope of this book, there are some aspects crucial for the present research that need to be highlighted.

Resilience is not intended as discrete step (Cuny 1983) implying some sort of outcomes, neither a quality (Patel and Nosal 2016). Resilience implies, in case of a disruption, that dysfunction is temporary, and system's resources can counteract the effects of the stressor so that pre-crisis functionality is restored. Rather than being a state, resilience is an ongoing process, where the system behaviour is observed not after the disruption, but while it is occurring understanding the process of change. Resilience is built and enhanced through resources, competences, knowledge, information, and as said it cannot be considered as a fixed quality of a system but rather as a process which takes place in time and learns from past experiences (Menoni 2020). "Spatial resilience" does not represent only the ability to recover from the disruption but also the ability to avoid or minimize the need to recover and the capacity to withstand unforeseen changes in a co-evaluative dimension (Brunetta and Caldarice 2020). Resilience can therefore be considered as a dynamic property that is constantly changing and indicates transformations of the system under analysis over time (Irwin et al. 2016), highlighting vulnerabilities, but also resources, internal dynamics, interdependences, and behavioural characteristics that can guide systems towards enhanced management in case of disruption while learning from past experiences. Most importantly, being resilience a time-dependent property, it can be either enhanced or reduced over time in function of development and redevelopment processes (Woods 2019).

The typical behaviour over time of a general system performance when a stressor is applied, is represented in Fig. 1.1.

When a disruption occurs, due to an applied stress, the general performance of a system described by the horizontal line (P_0), faces some alterations (ΔP) and precipitates. The recovery phase starts once the stressor ends acting on the performance of the system, till a new state of performance (P_f) is reached. Δt represents the response (from the disruption till the maximum loss of functionality/performance) and the

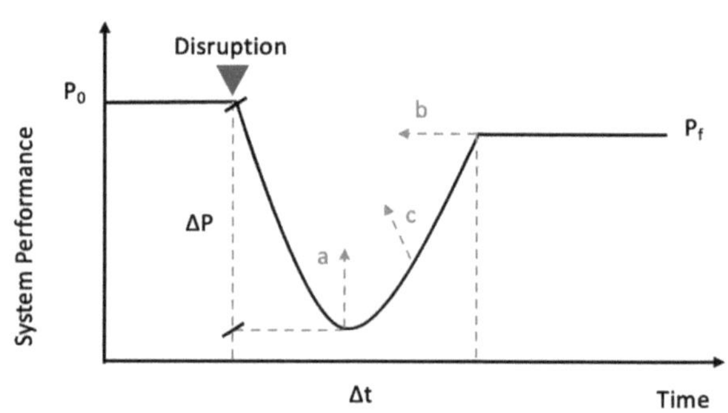

Fig. 1.1 Typical process of a general system performance changes after a disruption (Readapted from Fox-Lent and Linkov 2018)

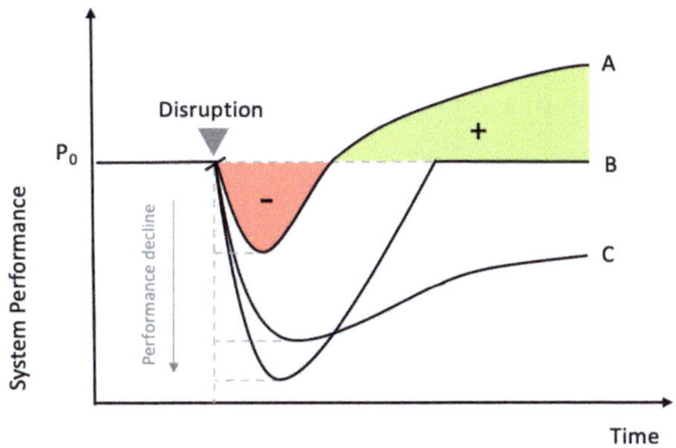

Fig. 1.2 Different system behaviours in case of a disruption

recovery (from the lowest level of performance to its new state) phases in terms of time spent to reach a new state of steadiness. It is in such Δt that resilience actively manifests. Specifically, a system can be considered resilient if the recovery curve (c), the magnitude of the disruption (a), and the recovery time (b) are successfully managed. The more flatten the curve, which means the performance of the system is maintained constant even when a disruption occurs, the more resilient a system is. The new level of performance (P_f) is a function of different factors, a resilient system recovers to the initial full capacity or even bounces forward to a higher level of functionality; on the contrary, a poor resilience management may inhibit the complete recovery of the system performance.

Figure 1.2 shows different systems' behaviours, the smaller the performance decline (red area) and the more rising the slope of the curve is (green area), then the more resilient is the system. Curve A shows the behaviour of a system able to enhance its performance after disruption. Curve B represents a system which returns to the previous level of performance. The last curve C represents a system that re-establishes a lower level of performance than the initial one.

Traditional risk analyses focus only on the performance decline (the ΔP represented in Fig. 1.1), describing the effects of a disruption on a system (Fox-Lent and Linkov 2018). However, given the complexity of nowadays societies and networks, understanding the capacity of systems to perform throughout disruptive events becomes essential. Yet, available resilience assessments do not contemplate the temporal component of event cycle, rather they propose methodologies very similar to an enhanced risk, or vulnerability, assessment (Fox-Lent and Linkov 2018; Bakkensen et al. 2016). As Fox-Lent and Linkov (2018) state, available approaches do not properly focus on reduction of the recovery time and/or on improving the shape of the recovery curve to re-establish an efficient performance state at an earlier point on the time axis after the threat has passed.

Resilience of a system plays a crucial role in different phases related to a disruptive event: (i) planning and preparation, (ii) absorption, (iii) recovery, and (iv) adaptation (Hollnagel et al. 2011; National Research Council 2012). It is crucial to evaluate the system performance in all the temporal phases with a strategic approach that

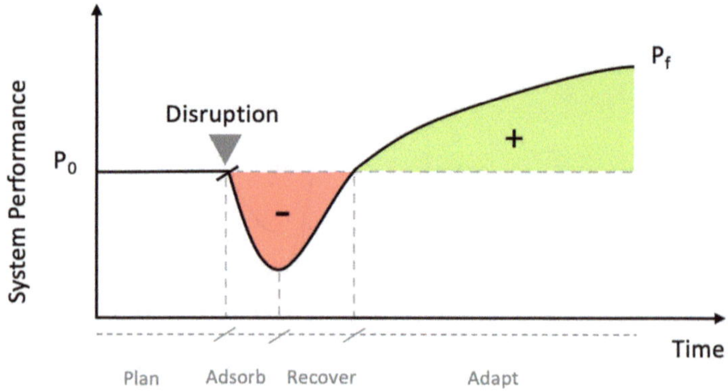

Fig. 1.3 Resilient system behaviour during plan, adsorb, recover and adapt phases (Readapted from Fox-Lent and Linkov 2018)

aims at generating systems able to minimize the negative impacts of adverse events improving their functionality by adapting to and learning from fundamental changes caused by the occurred stress (Linkov et al. 2014). Figure 1.3 comprises those phases and delivers a better idea of "process". In fact, resilience is considered as a dynamic process which builds on resources, competences, knowledge, information and lesson learned from past experiences, focusing on the behaviour of the system alongside the whole four phases and not limited to one of the latter. Figure 1.3 can be imagined as a single cycle of the whole process (see Fig. 2.1); the curve, that can assume different configurations as shown in Fig. 1.2, according to the actual performance of the system, repeats in time. Such curve varies in time according to the witnessed resilience of the system, which results in an improved or worsen performance.

1.2.2 Opportunities and Challenges of Shifting from Disaster Risk Management to Resilience Management

Shifting from disaster risk management to resilience management means catching up with the complexity that characterizes contemporary systems. Considering that future changes are hard to predict and the impact of disruptions are not always preventable, urban systems should learn how to cope and live with risks, which are not always well known.

The concept of resilience allows a proper conceptualization of cross-scale, cross-border and cross-sectoral interactions, while using different perspectives it allows to obtain innovative results which go beyond the resistance, robustness and return to the previous state concepts comprising opportunities of alternative futures and transformations of societal systems (Landstedt and Holmström 2007). Resilient approaches are operationally able to be adaptive and responsive and allow to constantly learn from success and failure in challenging and continuously changing systems (Brunetta and Caldarice 2020).

As Walker and Salt (2012) propose, shifting towards a "Resilience thinking", should appeal to anyone interested in dealing with risks in a complex world, as it offers constructive alternatives that create options rather than limiting them; redirecting towards the management of systems' capacities to cope with, adapt to, and shape change (Folke 2006).

However, embracing resilience management approach, nowadays, might be challenging. Being a novel approach, still not put in practice in a comprehensive way, it presents limitations that are not well known, as its related challenges, complexities and areas of improvements that must be still properly identified.

Whilst the risk management approach is far more consolidated in terms of available methodologies, operationalizing the conceptual framework of resilience management and finding indicators and scores to assess its level proved to be far more difficult. In addition, a gap between the more practical application performed by public administrations and the well refined theories and methodologies developed by researchers in the field is still rather large (Menoni 2020). Being the approach more recent and innovative, metrics to evaluate its efficiency and to allow comparisons within different strategies are still in the early stages of development (Linkov et al. 2013; Henry and Ramirez-Marquez 2012; Jones 2019). Additionally, tools and methods which have been developed insofar to assess resilience are very heterogeneous as both for inputs and outputs. Tools and methods make use of qualitative or quantitative data, manage them in different ways producing a variety of outputs from maps to tables, chart, diagrams (Linkov et al. 2018), requiring proper methodologies able to deal with such variety. Such methods must also account for disruptions that may be no longer restricted to any geographic vicinity due to high level of interconnectivity of modern society; the globalization and connectivity complicate the action background. The unpredictable nature of disruptions makes resilience measurements, monitoring and criteria identification difficult. Measuring resilience requires therefore methodologically different approaches (Béné 2013), entailing a wide range of steps ranging from the identification of appropriate definitions and frameworks, to the collection and use of appropriate data (Jones 2019). An additional significant gap is evident in the assessment of the reliability of resilience measures, when frameworks are operationalized in concrete applications (Saja et al. 2019). Advancing in this sense would allow to answer questions as "How resilient is the system? What can be done to enhance its resilience?". Being able to measure resilience is key to evaluate progress towards resilience. Metrics need to be developed following rational definitions in order to be integrable and aggregable to other metrics to cover the complexities of systems at hand (Ayyub 2015).

Despite the gaps and challenges, there are also opportunities for innovation in resilience measurement, where traditional approaches can be complemented and enhanced by new ones properly tailored, while making use of different available tools (Jones 2019).

Having good, reliable and comprehensive sets of data is a prerequisite for interpretation and for the achievement of proper and meaningful results and experimentation with innovative breakthroughs.

While risk approaches start form the consideration of hazards, often handled in disjunction between one type and another, rarely considered in concrete applications as interacting and enchained and where dynamism of vulnerability hardly is taken into account; resilience finds its roots in systems analysis discipline and therefore is, due to its nature, more complex.

Resilience recognizes that processes and dynamics that affect people and their territories are occurring across multiple spatial-scales and that they have feedbacks and recognizes links and interdependencies between components. Despite the challenges an advantage of resilience management approach is that puts together different actors, irrespectively of their backgrounds. Resilience approach creates "communication bridges" between disciplines and communities of practices, offering "common grounds" on which dialogue can be initiated. Due to its nature resilience leads the integration between spatial planning, disaster risk reduction and climate change adaptation, being central and of common interest to all three domains providing to those involved in the processes with opportunities to work together (Béné et al. 2012).

1.3 Damage and Loss Data to Reduce and Mitigate Future Risks

The relevance of damage and loss data as base elements for different applications to provide knowledge to a variety of stakeholders has become increasingly evident in the past years. De Groeve et al. (2013) were among the first to address the several uses of damage data. Menoni et al. (2017) highlighted the relevance of disaster damage and losses as a mean to improve risk modelling by uncovering the empirical features of hazard, exposure and vulnerabilities that generated such damage. The work of Ehrlich et al. (2017) provides recommendations for future loss data recording to be used to address disaster risk reduction as required for international reporting, and Marín Ferrer et al. 2018 highlighted the value of damage data as key element to support a variety of actions in relation to different EU policies and international initiatives to allow countries to respond to international agreements such as the Sendai Framework for Disaster Risk Reduction or the Agenda 2030 and so on.

Information and useful knowledge can be drawn from damage and loss data; in this context, detailed studies have been and are conducted in EU founded projects such as IDEA (Improving Damage assessments to Enhance cost–benefit Analyses), EDUCEN[4] (European Disasters in Urban Centers: a Culture Expert Network), LODE[5] (Loss Data Enhancement for DRR & CCA management),

[4] https://doi.org/10.3030/653874.

[5] https://www.lodeproject.polimi.it

AHEAD[6] (Advanced disaster damage and loss data information system for enhanced impact-based knowledge).

Damage and losses due to natural disaster can regard one or more exposed assets or multiple societal sectors ranging from lifelines to industries, agriculture, health-care system. Damage can be expressed as physical damage to assets and/or monetary costs of repair, as loss to individual economic sectors or to a given economy and society as a whole. In some cases damage cannot be quantified nor monetized. More specifically, following Menoni et al. (2016), damages can be considered as:

- Physical damage (damage to structures, assets, flora and fauna);
- Economic damage (estimated damage, unemployment, loss in salary);
- Loss data (evacuated people, injured, victims, psychological distress); and,
- Functional disruption (lack of services).

Such type of damages occurred due to natural hazards, in turn, can be generally classified, based on IDEA project and literature (Merz et al. 2010; Meyer et al. 2013; Van der Veen and Logtmeijer 2005), as:

- Damage to *tangible* objects and assets, meaning those for which a monetary assessment is generally easily obtained (buildings, assets, facilities);
- Damage to *intangibles*, meaning values for which monetization is either difficult or controversial (human life, monuments and historic heritage or environmental assets).

In addition, damage to both tangibles and intangibles can be once more classified in:

- *Direct*, meaning the immediate damages provoked directly by the event;
- *Indirect*, resulting from the direct damage or cascade effects due to the interdependency of communities' systems.

Direct damage generally occurs locally, indirect damage can develop over much larger time and space scales therefore, such type of damage is also identified as "second or higher order damage" (Rose 2004). Figure 1.4 illustrates typologies of direct and indirect damage and losses that can potentially impact different societal sectors.

Damage and loss data can reveal weakness of current risk management practices and identify possibilities for a change. Understanding the impact of disasters provides a profound knowledge that opens a window of opportunities for a range of stakeholders to work together on building safer and more sustainable communities. Specifically, they can support a variety of actions (Faiella 2020) as policy monitoring (construction of trends, observation of vulnerability and resilience fluctuations, forensic analysis); policy actions (compensation, legislation, investments, cost benefit analysis); prevention and preparedness (early warning systems, response

[6] https://civil-protection-knowledge-network.europa.eu/system/files/2025-03/ahead.pdf.

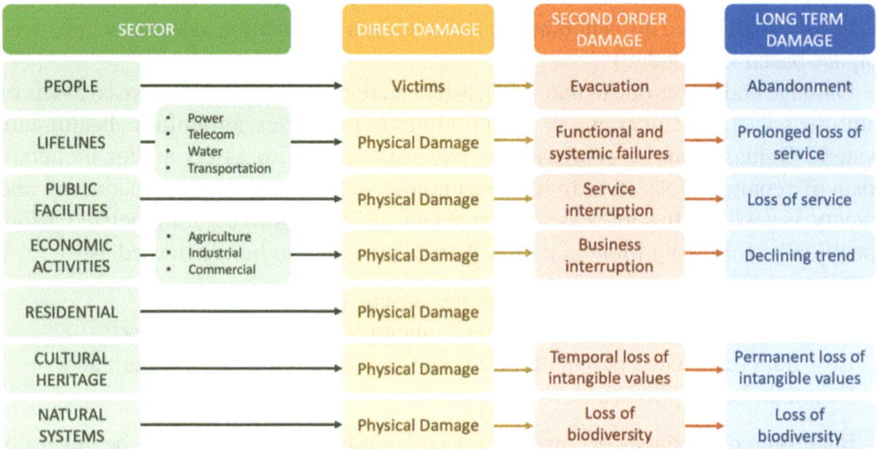

Fig. 1.4 Damage and loss framework for different societal sectors

actions, reconstruction process with a view to build back better, response); disasters' impact evaluation (loss accounting, systemic and cascading effects); and, risk modelling (deterministic and probabilistic).

The uses of post disaster damage data have been insofar initially considered mainly for recovery, reconstruction and prevention policies, while less attention has been posed on their relevance in the fields of emergency and crisis management where they may provide useful insight on how to improve preparedness by providing more suitable and focused response during an emergency. Specifically areas of application of damage and loss data can be schematized as follow:

- Loss Accounting: damage data collection and estimation serve the primary need related to the usability assessment of several assets ranging from residential buildings to critical facilities such as schools, hospitals and infrastructures. Also, primary damage data collection and estimation can assist in the declaration of state of emergency. Besides, insurance companies make use of the early estimate of the financial losses to draw an initial scenario of the damages. Finally, loss accounting is central to support the Sendai Indicators delineation.
- Disaster Forensic: forensic investigation allows to identify and analyse the causes of damages, it aims at digging into the underlying drivers, direct and root causes of the damages that have occurred to redress the various mitigation measures that should have prevented them. Among the latter early warning systems preparedness and capacities, land use and spatial planning.
- Needs Assessment: assessment of damage data is necessary for evaluating and prioritizing needs, intended as immediate necessary repairs, relocation, evacuation but also the needs arising on the longer time scale. Needs assessment is a practice that should follow a cyclic approach since needs vary over time along the disaster cycle (Solidarity Fund request, clearance of debris and special waste).

- Risk Modelling: deeper understanding of damages allows to better comprehend risk factors (hazard, exposure, vulnerability). Empirical damage and loss data recorded post-event can improve risk modelling (pre- and post-event) and can be used to validate and calibrate models.

The enhancement of one of those uses can support the improvement of the other since they are strictly related one to another (Walia et al. 2020): for instance forensic analysis, which supports the understanding of the role of hazard, exposure and vulnerability while analyses damages, returns crucial elements for an enhanced risk assessment, which in turn would support enhanced ex-ante modelling, emergency planning and so on. Therefore, empirical evidence of the impact that a disaster has on a territory, is a knowledge opportunity for several purposes. When a territory is affected by a disastrous event, physical assets are impacted, such as their functions and the lives of people. Damage dynamics are fundamental to be known to properly understand the risk drivers and draw lessons to support future actions and strategies, as it is proven by the application of the proposed methodology in this book showing the contribution that enhanced post-disaster damage data and analysis can make in terms of knowledge and information to improve the quality and comprehensiveness of—not only of pre-event measures in general—but specifically of spatial planning.

A robust inventory of damage and loss data would allow to understand till what extent disaster risk reduction is effective and efficient by monitoring the impacts of implemented programmes, for instance though cost–benefit analyses of disaster prevention and protection strategies. Cost–Benefit analysis has become an increasingly recommended tool to support decision-making for risk prevention and mitigation (Pesaro et al. 2018). It allows to assess the economic feasibility of risk management strategies by comparing the costs of mitigation measures with the potential benefits generated in terms of avoided damages by using as starting point empirical damage and losses, i.e. real observed impacts occurred. Empirical damage and loss data allow also to create a complete event scenario to understand the vulnerability of both the impacted area and community to support the delineation of enhanced risk assessments that would allow to address more efficient risk mitigation strategies. The SCENARIO project, founded under the VI Framework Programme (Support on Common European Strategy for Sustainable Natural and Induced Technological Hazards Mitigation 2006–2008), highlighted also the need to shift from "symptoms understanding" to "causes analysis" through a systemic approach. The degree of vulnerability of systems, assets or communities is a complex evidence to provide, and can be better identified in post-event conditions. In fact, vulnerability becomes evident only in the aftermath of an event when damage and loss are comprehended. Damage and losses are important data sources for vulnerability assessments. An accurate assessment of damages can support the development of post-disaster recovery programs, which should be delineated in time of peace, before the disaster strikes making use of knowledge from past events (Smith 2010).

1.3.1 Enhanced Damage and Loss Data to Improve Spatial Planning Base-Knowledge

Knowledge about the potential impact of natural events is necessary to achieve safer and more sustainable territories, decision and policy makers should be aware of existing risks and their potential impacts, but they also must understand how new spatial configurations and changes in land use can exacerbate (or mitigate) risks. Lack of or poor historic data limit the delineation of appropriate risk analysis to integrate in spatial planning (Burby et al. 2000). In addition, climate change impacts became increasingly an additional reminder of the solid connection between past and future challenges (Smith 2010). The availability of damage and loss data collected through a systematic collection process can help in the immediate aftermath of a disaster to identify priorities for intervention, during the emergency stage, but also in the longer term to mitigate future risks through efficient spatial planning, which reduces proactively exposure and vulnerability, while increasing resilience. Suitable results can be obtained only by adopting a more integrative approach that seeks solutions for both natural hazards and climate change considered together in a comprehensive framework of sustainability (Menoni and Margottini 2011).

It is licit to question how post-event damage and loss data (which illustrate dynamics of past occurrences) can be linked to spatial planning (which directs the present configurations into the future). Damage & loss data can "describe", on the basis of what has occurred in the past, how future may unfold allowing to compare the potential consequences on different future contexts to support spatial planning aimed at reducing risks while increasing resilience, by improving skills and capacities to anticipate possible scenarios. The territory that we observe today is the result of a slow process, choices and decisions coming from the past often not coherent between each other. Decision that could seem coherent in the short term have proved to be contradictory after a period of time. Therefore, learning from the past and observing current configurations may guide in the construction of valuable scenarios, as highlighted by Secchi (2011) in "Prima lezione di urbanistica", according to whom scenarios are the link that "sews and binds" the patterns of the past with the visions for the future. Therefore, it is necessary to enhance awareness and consciousness that damages occurred due to hazardous event can generate and enhance a better configuration of functions for the future. Constructing knowledge learning from past can motivate governments at all levels to focus on past disasters to prevent future losses (Donahue and Tuohy 2006) and enhance resilience in the face current and future risks.

Analysis of past disasters provide a unique opportunity to address changes (Smith 2010) by understanding damage dynamics, since it allows to uncover origins and causes which can be amended for future configurations. Such understanding is the precondition to improve strategies which aim at achieving not only disaster risk reduction, but territorial systems which embrace resilience. Damage and loss data contribute to the identification of the causes of disasters impacts becoming an added value for spatial planning. Knowledge of post-event provides evidence can guide planners towards resilient configurations, considering that "resilience relies on resources, capacities, ability to learn from the experience lived through a disaster" (Menoni 2020). Damage and loss data, if properly and systematically collected, could create a relevant repository of information to detect the most damaged zones and sectors to identify most vulnerable areas and to understand behavioural patterns of systems when affected by a disruption. This approach becomes an opportunity for the enrichment of the base-knowledge needed to implement a proactive and adaptive spatial planning which integrates concepts of disaster risk reduction framed within the concept of resilience.

However, the current settings of organized and structured damage and loss data collection and storage are still unsatisfactory (Faiella et al. 2022), as proven by the lack of basic figures of disaster losses at the scale, granularity and level of detail needed for most of the applications described above; although there is an important international endeavour aimed at moving forward current limitations, inconsistencies, and gaps of existing datasets and cross-sector data collection for the future.

References

J. Adams, *Risk* (UCL Press, London, 1995)

D.E. Alexander, Resilience and disaster risk reduction: an etymological journey. Nat. Hazard. **13**(11), 2707–2716 (2013)

B.M. Ayyub, Practical resilience metrics for planning, design, and decision making. ASCE-ASME J. Risk Uncertain. Eng. Syst. Part a: Civil Eng. **1**(3), 04015008 (2015)

L.A. Bakkensen, C. Fox-Lent, L.K. Read, I. Linkov, Validating resilience and vulnerability indices in the context of natural disasters. Risk anal. **37**(5), 982–1004 (2016)

C. Béné, Towards a quantifiable measure of resilience. IDS Work. Pap. **2013**(434), 1–27 (2013)

C. Béné, L. Mehta, G. McGranahan, T. Cannon, J. Gupte, T. Tanner, Resilience as a policy narrative: potentials and limits in the context of urban planning. Clim. Dev. **10**(2), 116–133 (2018)

C. Béné, R.G. Wood, A. Newsham, M. Davies, Resilience: new utopia or new tyranny? Reflection about the potentials and limits of the concept of resilience in relation to vulnerability reduction programmes. IDS Work. Pap. **2012**(405), 1–61 (2012)

P. Bolton, S. Heikkala, M.M. Green, P. May, *Land Use Planning for Earthquake Hazard Mitigation: A Handbook for Planners*. Special Publication, vol. 14 (Institute of Behavioral Sciences, University of Colorado, Boulder, 1986)

M.P. Boni, A. Faiella, V. Gazzola, F. Pergalani, A multi-hazard and multi-risk assessment methodological approach to support civil protection planning in wide areas. Int. J. Disast. Risk Reduct. **119**, 105343 (2025)

G. Brunetta, O. Caldarice, Spatial resilience in planning: meanings, challenges, and perspectives for urban transition. Sustain. Cities Commun. 628–640 (2020)

R.J. Burby (ed.), *Cooperating with Nature: Confronting Natural Hazards with Land-Use Planning for Sustainable Communities* (Joseph Henry Press, 1998)

R.J. Burby, R.E. Deyle, D.R. Godschalk, R.B. Olshansky, Creating hazard resilient communities through land-use planning. Nat. Hazard. Rev. **1**(2), 99–106 (2000)

D. Chandler, *Resilience: The Governance of Complexity* (Routledge, 2014)

COM/2021/82. Communication from the commission to the European parliament, the council, the European economic and social committee and the committee of the regions empty, in *Forging a Climate-Resilient Europe—the New EU Strategy on Adaptation to Climate Change*. Available at https://eur-lex.europa.eu/legal-content/EN/TXT/?uri=COM:2021:82:FIN

D. Coppola, *Introduction to International Disaster Management* (Elsevier, 2006)
F.C. Cuny, *Disasters and Development* (Oxford University Press, New York, 1983)
R.J. Dawson, T. Ball, J. Werritty, A. Werritty, J.W. Hall, N. Roche, Assessing the effectiveness of non-structural flood management measures in the Thames Estuary under conditions of socio-economic and environmental change. Glob. Environ. Chang. **21**(2), 628–646 (2011)
A. Dazé, A. Terton, M. Maass, *Coordinating Climate-Resilient Development Alignment to Advance Climate-Resilient Development. Overview Brief 1: Introduction to Alignment*, NAP Global Network (2018). http://napglobalnetwork.org/wp-content/uploads/2018/08/napgn-en-2018-alignment-to-advance-climate-resilient-development-overview-brief.pdf
T. De Groeve, K. Poljansek, D. Ehrlich, *Recording Disaster Losses: Recommendation for a European Approach*, EUR 26111 ENPublications Office of the European Union, Luxembourg, JRC83743 (2013)
A.K. Donahue, R. V. Tuohy, Lessons We don't Learn: A Study of the Lessons of Disasters, Why we Repeat Them, and How we can Learn Them (2006)
R.R. Dynes, Relevance for policy and practice, in *Improving Earthquake Mitigation: Report to Congress...* (The Office, 1993), p. 67
D. Ehrlich, C. Corbane, T. De Groeve, Technical recommendations for standardizing loss data, in *Flood Damage Survey and Assessment: New insights from Research and Practice*. ed. by D. Molinari, F. Ballio, S. Menoni (AGU (American Geophysical Union) series, Wile, Washington, D.C., 2017), pp.17–29
J.F. Esteban, B. Izquierdo, J. Lopez, D. Molinari, S. Menoni, A. De Roo, ... G. Eftichidis, Current mitigation practices in the EU, in *Inside Risk: a Strategy for Sustainable Risk Mitigation* (Springer, Milano. Kellermann, 2015)
A. Faiella, Agriculture damage data collection: a model for reconstructing comprehensive damage dynamics. Progr. Disast. Sci. **7**, 100112 (2020)
A. Faiella, S. Menoni, M.P. Boni, M. Panoutsopoulou, T. Thoma, S. Salari, N. Rueda, Enabling knowledge through structured disaster damage and loss data management system. Sustainability **14**(10), 6187 (2022)
C. Folke, Resilience: the emergence of a perspective for social–ecological systems analyses. Glob. Environ. Chang. **16**(3), 253–267 (2006)
C. Fox-Lent, I. Linkov, Resilience matrix for comprehensive urban resilience planning. Resilience-Oriented Urban Planning: Theor. Empirical Insights 29–47 (2018)
S.Fuchs, T. Thaler (eds.), *Vulnerability and Resilience to Natural Hazards* (Cambridge University Press, 2018)
A. Galderisi, S. Menoni, Risk prevention and urban planning. Urbanistica J **134**, 20–23 (2007)
A. Galderisi, A.S. Profice, Land use planning, sustainability and risk mitigation: towards an integration, in *Land Use Planning and Environment*, ed. by S. Aubrecht, S. Freire, K. Steinnocher (Nova Science Publisher, Inc., New York, 2012)
V. Gazzola, in *Uso del suolo e riduzione del rischio naturale in ambito urbano.* Maggioli Editore (2019)
GCSA (2022). Improving EU crisis prevention, preparedness, response and resilience. Group of Chief Scientific Advisors, Scientific Opinion No.13, November 2022 GCSA, in *Improving EU crisis prevention, preparedness, response and resilience. Group of Chief Scientific Advisors, Scientific Opinion No.13, November 2022.* European Commission, Directorate-General for Research and Innovation (2022). https://doi.org/10.2777/517560 ISBN 978-92-76-53947-6
D. Godschalk, T. Beatley, P. Berke, D. Brower, E.J. Kaiser, in *Natural Hazard Mitigation: Recasting Disaster Policy and Planning* (Island Press, 1998)
D. Henry, J.E. Ramirez-Marquez, Generic metrics and quantitative approaches for system resilience as a function of time. Reliab. Eng. Syst. Saf. **99**, 114–122 (2012)
E. Hollnagel, J. Paries, D. Woods, J. Wreathall, *Resilience Engineering in Practice: A Guidebook* (Ashgate, United Kingdom, 2011)
L.D. Hopkins, *Urban Development: The Logic of Making Plans* (Island Press, D.C., 2001)

M. Huxley, A. Inch, Urban planning, in *International Encyclopedia of Human Geography (Second Edition)* (2020), pp. 87–92

IPCC. Global warming of 1.5°C An IPCC Special Report on the impacts of global warming of 1.5°C above pre-industrial levels and related global greenhouse gas emission pathways, in *the Context of Strengthening the Global Response to the Threat of Climate Change, Sustainable Development, and Efforts to Eradicate Poverty, Intergovernmental Panel on Climate Change (IPCC)* (2018). https://www.ipcc.ch/sr15/

S. Irwin, A. Schardong, S.P. Simonovic, N. Nirupama, ResilSIM—a decision support tool for estimating resilience of urban systems. Water **8**(9), 377 (2016)

L. Jones, Resilience isn't the same for all: comparing subjective and objective approaches to resilience measurement. Wiley Interdisciplin. Rev. Clim. Change **10**(1), e552 (2019)

M. Kammerbauer, in *Planning Urban Disaster Recovery*. VDG Weimar (2013)

D. King, Y. Gurtner, A. Firdaus, S. Harwood, A. Cottrell, Land use planning for disaster risk reduction and climate change adaptation: operationalizing policy and legislation at local levels. Int. J. Disast. Resilien. Built Environ. (2016)

R.J. Klein, R.J. Nicholls, F. Thomalla, Resilience to natural hazards: how useful is this concept? Glob. Environ. Change Part b: Environ. Hazards **5**(1), 35–45 (2003)

J. Landstedt, P. Holmström, Electric power systems blackouts and the rescue services: the case of Finland, in *Emergency Services College of Finland and State Provincial Office of Western Finland* (2007)

I. Linkov, T. Bridges, F. Creutzig, J. Decker, C. Fox-Lent, W. Kröger, ... T. Thiel-Clemen, Changing the resilience paradigm. Nat. Clim. Change **4**(6), 407–409 (2014)

I. Linkov, C. Fox-Lent, L. Read, C.R. Allen, J.C. Arnott, E. Bellini, ... D. Woods. Tiered approach to resilience assessment. Risk Anal. **38**(9), 1772–1780 (2018)

I. Linkov, D.A. Eisenberg, K. Plourde, T.P. Seager, J. Allen, A. Kott, Resilience metrics for cyber systems. Environ. Syst. Decis. **33**(4), 471–476 (2013)

K. MacAskill, P. Guthrie, Multiple interpretations of resilience in disaster risk management. Procedia Econ. Financ. **18**, 667–674 (2014)

S.B. Manyena, The concept of resilience revisited. Disasters **30**(4), 433–450 (2006)

A. March, M. Kornakova (eds.), in *Urban Planning for Disaster Recovery* (Butterworth-Heinemann, 2017)

A. March, J. Leon, Urban planning for disaster risk reduction: Establishing 2nd wave criteria, in *State of Australian Cities*, ed. by K. Ruming, B. Randolph, N. Gurran (Sydney, NSW: State of Australian Cities Research Network, 2013)

M. Marín Ferrer, A. Do Ó, K. Poljanšek, A. Casajus Vallés, in *Disaster Damage and Loss Data for Policy: Pre- and post-event damage assessment and collection of data for evidence-based policies, JRC Science for Policy Reports, Publications Office of the European Union, Luxembourg* (2018)

S. McManus, E. Seville, D. Brunsden, J. Vargo, in *Resilience Management: A Framework for Assessing and Improving the Resilience of Organisations* (2007)

S. Menoni, Different preventive strategies require diversified risk assessment models. Environ. Secur. Environ. Manag. Role Risk Assess. 211–221 (2006)

S. Menoni, Pianificazione urbanistica e territoriale in aree soggette a rischi naturali: limiti e opportunità. Sentieri Urbani **7**, 20–27 (2012)

S. Menoni, Risk mitigation and resilience of human settlement, in *Risk and Resilience: Socio-Spatial and Environmental Challenges*. ed. by A. Balducci, D. Chiffi, F. Curci (Springer International Publishing, 2020), pp.27–47

S. Menoni, G. Pesaro, Is relocation a good answer to prevent risk? Criteria to help decision makers choose candidates for relocation in areas exposed to high hydrogeological hazards. Disaster Prevention and Management: An Int. J. **17**(1), 33–53 (2008)

S. Menoni, C. Margottini (eds.), in *Inside Risk: A Strategy for Sustainable Risk Mitigation* (Springer Science & Business Media, 2011)

S. Menoni, A. Galderisi, D. Carrion, C. Gerosa, Cross-sectoral and multilevel dimensions of risk and resilience management in urban areas enabled by geospatial data processing. Sustainability **16**(19), 8712 (2024)

S. Menoni, F. Atun, D. Molinari, G. Minucci, N. Berni, Defining complete post flood scenarios to support risk mitigation strategies, in *Flood Damage Survey and Assessment: New Insights from Research and Practice; AGU (American Geophysical Union), Series*, eds. by D. Molinari, F. Ballio, S. Menoni, (Hoboken, NJ, USA, Wiley, 2017)

S. Menoni, D. Molinari, F. Ballio, G. Minucci, O. Mejri, F. Atun, C. Pandolfo, Flood damage: a model for consistent, complete and multipurpose scenarios. Nat. Hazards Earth Syst. Sci. **16**(12), 2783–2797 (2016)

B. Merz, H. Kreibich, R. Schwarze, A. Thieken, Review article Assessment of economic flood damage. Nat. Hazards Earth Syst. Sci. **10**(8), 1697–1724 (2010)

V. Meyer, N. Becker, V. Markantonis, R. Schwarze, J.C. van den Bergh, L.M. Bouwer, C. Viavattene, Assessing the costs of natural hazards–state of the art and knowledge gaps. Nat. Hazards Earth Syst. Sci. **13**(5), 1351–1373 (2013)

S. Mohammadi, S. De Angeli, G. Boni, F. Pirlone, S. Cattari, Current approaches and critical issues in multi-risk recovery planning of urban areas exposed to natural hazards. Egusphere **2023**, 1–39 (2023)

National Research Council, in *Disaster Resilience: A National Imperative* (The U.S. National Academy of Sciences: The National Academies Press: Washington, DC., 2012)

F.H. Norris, S.P. Stevens, B. Pfefferbaum, K.F. Wyche, R.L. Pfefferbaum, Community resilience as a metaphor, theory, set of capacities, and strategy for disaster readiness. Am. J. Community Psychol. **41**, 127–150 (2008)

M.J. Nowak, R. Monteiro, J. Olcina-Cantos, D.G. Vagiona, Spatial planning response to the challenges of climate change adaptation: an analysis of selected instruments and good practices in Europe. Sustainability **15**(13), 10431 (2023)

OECD, Common ground between the Paris agreement and the sendai framework: climate change adaptation and disaster risk reduction. OECD Publishing, Paris, (2020). https://doi.org/10.1787/3edc8d09-en

N. Okada, Urban diagnosis and integrated disaster risk management. J. Nat. Dis. Sci. **26**(2), 49–54 (2004)

A. Oliver-Smith, Anthropological research on hazards and disasters. Annu. Rev. Anthropol. **25**(1), 303–328 (1996)

J. Park, T.P. Seager, P.S.C. Rao, M. Convertino, I. Linkov, Integrating risk and resilience approaches to catastrophe management in engineering systems. Risk Anal. **33**(3), 356–367 (2013)

R. Patel, L. Nosal, *Defining the Resilient City* (United Nations University Centre for Policy Research, New York, 2016)

R.W. Perry, M.K. Lindell, Earthquake planning for government continuity. Environ. Manage. **21**, 89–96 (1997)

G. Pesaro, M.T. Mendoza, G. Minucci, S. Menoni, Cost-benefit analysis for non-structural flood risk mitigation measures: Insights and lessons learnt from a real case study, in *Safety and Reliability–Safe Societies in a Changing World* (CRC Press, 2018), pp. 109–118

A. Rose, Economic principles, issues, and research priorities in hazard loss estimation. Model. Spat. Econ. Impacts Disast. 13–36 (2004)

P.H. Rossi, J.D. Wright, E. Weber-Burdin, in *Natural Hazards and Public Choice: The State and Local Politics of Hazard Mitigation* (Elsevier, 2013)

F. Roy, Y. Ferland, Land-use planning for disaster risk management. Land Ten J. (1) (2015)

K. Rus, V. Kilar, D. Koren, Resilience assessment of complex urban systems to natural disasters: a new literature review. Int. J. Disast. Risk Reduct. **31**, 311–330 (2018)

H. Rutherford Planning adjustment in historical perspective, in *Cooperating with Nature: Confronting Natural Hazards With Land-Use Planning for Sustainable Communities*, ed. by R. J. Burby (Joseph Henry Press, 1998)

A.A. Saja, A. Goonetilleke, M. Teo, A.M. Ziyath, A critical review of social resilience assessment frameworks in disaster management. Int. J. Disast. Risk Reduct. **35**, 101096 (2019)
B. Secchi, in *Prima lezione di urbanistica* (Gius. Laterza & Figli Spa, 2011)
A. Sharifi, Y. Yamagata, Resilience-oriented urban planning, in *Resilience-Oriented Urban Planning: Theoretical and Empirical Insights* (2018), pp. 3–27
G. Smith, Lessons from the United States: planning for post-disaster recovery and reconstruction. Austral. J. Disast. Trauma Stud. **1**, 1–8 (2010)
The European Green Deal, COM/2019/640, in *Communication from the Commission. To the European Parliament, the European Council, the council, the European Economic and Social Committee and the Committee of the Regions*
The Territorial Agenda 2030, in *A Future for all places, TAEU 2030* (2020). Available at https://territorialagenda.eu/wp-content/uploads/TA2030_jun2021_en.pdf
UN Habitat, New urban agenda: Quito declaration on sustainable cities and human settlements for all, in *Quito UN Habitat* (2016)
UNDRO, *Natural Disasters and Vulnerability analysis. Report of Expert Group Meeting, July 1979* (Office of the United Nations Disaster Relief Co-ordinator, Geneva, 1979)
UNDRR, *United Nations Office for Disaster Risk Reduction.* Report of the open-ended intergovernmental expert working group on indicators and terminology relating to disaster risk reduction (2017). Available at https://www.undrr.org/publication/report-open-ended-intergovernmental-expert-working-group-indicators-and-terminology
UNISDR, in *The United Nations office for Disaster Risk reduction. Making Cities Resilient Report 2012* (2012). Available at: https://www.preventionweb.net/files/28240_rcreport.pdf
UNISDR, Reading the Sendai Framework for Disaster Risk Reduction 2015–2030, in *Support of the implementation of the Sendai Framework for Disaster Risk Reduction.* (Switzerland, Geneva, 2015), pp.2015–2030
D. Van den Hoek, T. Spit, T. Hartmann, Certain flexibilities in land-use plans towards a method for assessing flexibility. Land Use Policy **94**, 104497 (2020)
A.V.D. Veen, C. Logtmeijer, Economic hotspots: visualizing vulnerability to flooding. Nat. Hazards **36**(1), 65–80 (2005)
A. Walia, S. Menoni, M.F. Dell'Aringa, M. Bakon, M. Rovnak, K. Milenov, A. Dyulgerov, R. Radkov, L. Posthuma, T. Shopov, K. Markov, C. van de Guchte, M.A. Altamirano, D. Wagenaar, Methodologies for disaster impact assessment, in: *Science for Disaster Risk Management 2020: acting today, protecting tomorrow, EUR 30183 EN, Publications Office of the European Union, Luxembourg*, eds. A. Casajus Valles, M. Marin Ferrer, K. Poljanšek, I. Clark, ISBN 978–92–76–18182–8, https://doi.org/10.2760/571085, JRC114026
B. Walker, D. Salt, *Resilience Thinking: Sustaining Ecosystems and People in a Changing World* (Island press, 2012)
C. Wamsler, Mainstreaming risk reduction in urban planning and housing: a challenge for international aid organisations. Disasters **30**(2), 151–177 (2006)
C. Wamsler, Planning ahead: adapting settlements before disaster strike, in *Hazards and the Built Environment*, ed. by L. Bosher, (Taylor & Francis, London, 2008)
M. Wegener, GIS and spatial planning. Environ. Plann. b. Plann. des. **25**(7), 48–52 (1998)
G.F. White, *Human Adjustment to Floods: A Geographical Approach to the Flood Problem in the United States* (Department of Geography, University of Chicago, Chicago, 1945)
D.D. Woods, Essentials of resilience, revisited, in *Handbook on Resilience of Socio-Technical Systems* (2019), pp. 52–65
T. Yoshida, Y. Yamagata, S. Chang, V. de Gooyert, H. Seya, D. Murakami, ... G. Voulgaris, Spatial modeling and design of smart communities, in *Urban Systems Design* (Elsevier, 2020), pp. 199–255

Chapter 2
Stress Testing Urban Planning for Resilience: A Methodology

Abstract In this chapter the methodology that has been designed, developed and applied to a case study is explained and the rationale behind choices on selected approach, modifications to existing frameworks and tools are discussed. First an outline on how the concept of resilience applied to cities and to spatial planning is briefly recalled, highlighting the need to consider its multiscale temporal and spatial character. To do so two methods have been used and blended together to operationalize the evaluation of urban resilience: the resilience matrix and the stress testing. In the last part of the chapter the operational steps to apply the replicable methodology to different case studies are presented.

2.1 Boundaries Definition: Understanding of Urban Resilience

The proposed approach does not delimit a spatial scale of action, but rather a dimension. Methodologically, this work focuses on the spatial dimension of systems. More specifically, the present research explores the behaviour of urban spaces and their communities, intended as a system, in case of a disruption. Cities are complex spaces, strategic nodes in which social, political and economic organizations interact through different connections and at multiple levels with both its assets and functions such as buildings, infrastructures, production and services (Menoni and Atun 2017). Even though this work focuses on cities, the multiple scales and levels through which the latter are interconnected with other cities, with their region, nation and above cannot be neglected. As correctly put by Wilbanks and Kates (1999), analysis by scale obeys to a key requirement: put localized observations into a reference context. Considering only local aspects is actually impossible as nowadays local, regional, national and global scales are interconnected, though in different ways for metropolitan and central areas and for small-medium towns and marginal and non-central areas. The different scales are in tension with each other, and this tension owes a lot to the type of existing networks, both physical and non-physical, economic, social, and cultural (Root and Schneider 1995). Spatial scale is therefore a key concept in planning. It addresses the

A. Faiella, *Learning from Disasters to Enhance Resilience of Urban Complex Systems*, PoliMI SpringerBriefs, https://doi.org/10.1007/978-3-032-07356-3_2

fact that decisions and plans are often developed at different «scales», encompassing the «level» at which one is operating (local, regional, national, global) as well as the «administrative» and political levels of government that are implied and responsible for decisions and their implementation (municipality, county or province, region/department, nation, European space). Similarly, also natural phenomena impacting cities develop at different scales, depending on their intrinsic characteristics. A landslide can be considered a rather local occurrence, yet if several landslides are triggered at the same time by a large storm they may turn into a regional hazard. Other hazards, such as floods, can be comprehended and modelled only considering the entire catchment, within which cities represent only a section of the entire river basin. This fact points at one relevant difficulty in integrating hazard with exposure and vulnerability analysis: the scale at which phenomena and cities function do not always coincide. According to Root and Schneider (1995) in fact, «the scale at which different research disciplines operate make multidisciplinary connection difficult and necessitate devising methods for bridging scale gaps». This is a challenge that this research is facing and for which dedicated tools have been deployed to manage the large number of different data and information related to a multiplicity of processes and phenomena (both urban, natural and related to climate change) in a manner that allows deeper understanding of exposure and vulnerability to propose suitable measures to mitigate risks and enhance resilience.

Cities are considered here as a system of components and as a system of systems on which it relies to function and continue to develop (Pesaro et al. 2018). System's components are elements, objects, functions, services and activities, deployed within built and open environments. Such components represent the assets, or better the values, exposed to the threatening hazards. Their intrinsic features, instead, define the vulnerability of systems' components. The way systems are interconnected shapes the vulnerability (or on the opposite the strengths) of urban areas to physical and systemic disruptions. Urban systems are multipart configurations characterized by several systems which through interactions enable an appropriate functioning of the whole system. Economic, social, infrastructural, educational, trading systems interact in complex ways between each other and are in their turn constituted by the components described above. Despite such complexity being acknowledged by several authors (Pesaro et al. 2018; Bettencourt 2015; Allen 1997), few studies approach urban systems in their totality (Koren et al. 2017).

In terms of temporal scale, the boundaries are set by the concept of resilience. Our approach considers resilience with respect to different temporal phases, resilience is a process that has to be ongoing and enhanced before, during and after a disruption occurs (Pursiainen et al. 2016). Specifically, it is here intended as represented in Fig. 2.1 where t_i can be considered as a potential current status that has learned from the previous one occurred at time t_{i-1} and which consequently can have a positive effect on the final resilience of the system in future, as represented in time t_{i+1}. Specifically, resilience on the one hand is considered as (i) a process through which systems learn from the past to address the future, as represented in the figure where the configuration at time t_i is a consequence of the previous configuration at time t_{i-1}. On the other it is considered as (ii) operating across different phases of a

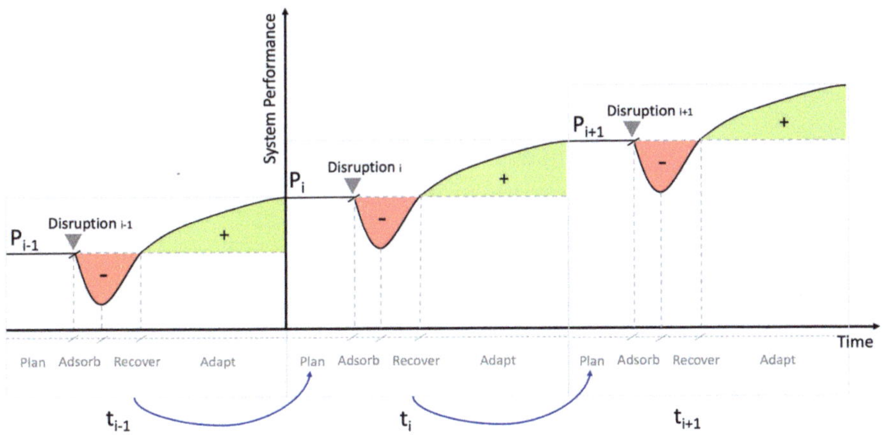

Fig. 2.1 Temporal scale set by the concept of resilience

disaster/disruption, namely across planning and preparation, absorption, recovery, and adaptation. It is in the planning phase that what has been learned from the past (blue arrows) can shape better preparation, absorption, recovery, and adaptation through a cyclic approach leading to enhanced system performance as P_i and P_{i+1} when compared with P_{i-1}.

2.2 Methodological Outline of Selected Frameworks and Tools

The research outline is conceptually represented in Fig. 2.2. The seven grey rectangles represent the main interconnected research segments implemented in different phases, namely: (1) analysis of the relationship between Disaster Risk Reduction and Spatial Planning to understand how they are addressed by research and policies and to understand their integration status and "dialogue" conceptually and operationally; (2) conceptual exploration of meaning of data and information; (3) understanding of the potential of damage and loss data to generate and integrate knowledge to support Spatial Planning. The exploration of these three segments is allocated in the first desk-based research phase which allowed to define a system thinking approach structuring the fourth research segment: (4) definition of a renewed approach which considers systematically Cities, Climate Change, Disaster Risk Reduction and the shift towards Resilience Management as represented in the blue boxes. Following the previous phases, data collection and the case study analysis were carried out. Finally, (5) in the yellow boxes, the comprehensive methodology using results of the damage and loss data analysis as key elements to feed the stress testing and of the resilience matrix has been elaborated—and applied in the case study (6). The results

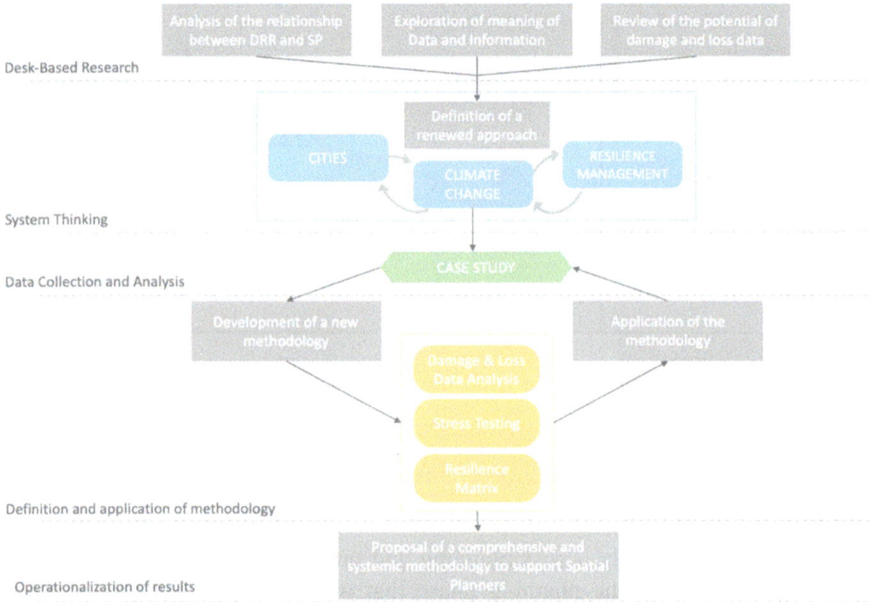

Fig. 2.2 Conceptual representation of the research outline

have been subsequently operationalized through (7) the proposal of a comprehensive and systemic methodology to support spatial planners.

The use of damage and loss data within a resilience framework combining stress testing and resilience matrix to support a proactive spatial planning required to define a precise architecture for the research. An exploratory approach guided the activities in order to identify the variables needed to reach the research objective, which descend from the three main pillars on which the research gets shaped: (i) data science, (ii) spatial planning; and (iii) resilience management.

In the following sections the tools that have been used to operationalize the stress testing through resilience matrices as a support to spatial planning are described in detail.

2.2.1 A Revised Resilience Matrix

The resilience matrix is a flexible methodology, proposed by Linkov et al. (2013a, b), which makes use of qualitative and quantitative data and information to evaluate the resilience level of a system. The study proposes that, physical (facilities, equipment, system states and capabilities), information (creation/collection, manipulation/management, and storage of data), cognitive (understanding, notions), and social (interaction, collaboration between stakeholders and entities) domains of a system should be assessed jointly rather than separately as often seen in academic literature. Those domains find a correlation within the event management cycle which considers planning and preparation, absorption, recovery, and adaptation as defined by the National Academy of Sciences (2012). The novelty of such approach lies in the contemplation of the system behaviour performance in function of the stress. The plan and adsorb phases need to be properly managed before and as soon as the threat begins, while recovery and adaptation allow to focus on the system after the occurrence so that efforts can be properly addressed to re-establish an optimum level of performance in a quick and efficient manner so the system can adjust and increase robustness to better handle future disruptions (Linkov et al. 2019).

The original resilience matrix, has been developed and applied for cyber systems, energy, engineering, and ecology (Eisenberg et al. 2014; Linkov et al. 2013a, b; Roege et al. 2014). In Fox-Lent et al. (2015) have adopted the framework for the resilience assessment of the Rockaway Peninsula community in New York, an area of relevance in terms of flood risk management, navigation, and ecological restoration in addition to being an area of interest following Hurricane Sandy. The assessment regarded coastal storm threats, including hurricanes using data collected after Hurricane Sandy. The final results, for instance, showed that the critical function of the housing, presented a better behaviour in the prepare and adsorb phases rather than recover and adapt ones; an expected outcome, as stated in the work, considering that for all similar contexts funding become available to communities immediately after the disaster has struck so money are used to return to the previous state rather than investing on new capacities. The study revealed opportunities to prioritize investments and collaborate among different stakeholders for resilience enhancement of some features of the system analysed to properly plan the management of adverse events. Successively, in Fox-Lent and Linkov (2018) showed an application on the case study of Mobile Bay region (Alabama) which in 1998 has been affected by a category 2 hurricane, named Georges, that provoked both inundation and coastal flooding due to the heavy rainfall and strong winds. The assessment was aimed at enhancing resilience and protection from coastal flooding from both tide and storm surge. The work entailed knowledge sharing among stakeholders and collection of available historical data, focused on housing/shelter, shipping, tourism, and the bay ecosystem. The overall behaviour showed a better capacity in the information and cognitive domains than in the physical and social domains. Observation demonstrated that the tourism industry has strong capacity for the "prepare" phase which, however, doesn't lead to improved capacity to absorb. Shipping and port related activities showed resilient capacities in the aftermath of the disruption but a lower adaptation profile. Similar for ecosystem, where no measures have been taken to prevent or minimize damage. According to the results in the cells for each analysed function, proposals have been made to support urban resilience planning such as: building code improvements and enforcement for coastal structures, especially on the barrier islands; use natural revetment along the bay to mitigate erosion, arrange a link with expert contractors in coastal storm damage mitigation for local business to access when necessary; enhance natural drainage by reducing impervious surfaces in new developments; enhance awareness on ecosystems.

		Stages of disaster management			
Domains of complex systems	RESILIENCE MATRIX	Prepare	Adsorb	Recover	Adapt
	Information				
	Physical				
	Economic				
	Social				

Fig. 2.3 Structure of the (revised) resilience matrix

Considering the potential of the resilience matrix to support planning processes, we propose a revised resilience matrix (see Fig. 2.3) largely built upon the matrix proposed by Linkov et al. (2013a, b), where the domains of the system have been enriched with "economic domain", whilst "information and cognitive domains" have been integrated together, as following:

- PHYSICAL DOMAIN where the event and responses occur. It includes the components that are directly impacted by the event and need to be brought back to full efficiency and functionality in order to have a resilient behaviour.
- INFORMATION and COGNITIVE DOMAIN where knowledge and data exists, are processed and shared to become to have a meaning. This domain is related to general data, information about the development and changes of the other domains, and assessment of the degree to which plans and strategies exist. The objectives of resilience in this domain are to prepare information for potential stresses and assure that the system will react efficiently after the occurrence, this components need to be properly addressed since they are crucial for a proper implementation of resilience operations.
- ECONOMIC DOMAIN in which the impact on the economic sphere of the system is considered. It is essential to understand the extent and the implication of the event over the economic cycle of the system, in order to evaluate its resilience capacity.
- SOCIAL DOMAIN considers the role of human decision making, it is characterised by interactions and collaboration between and within entities involved (community, stakeholders, decision makers). The information domain plays a crucial role regarding the whole functioning, information needs to be handled, understood, interpreted and trusted properly by all parts.

Those domains are then combined with the four phases related to the occurrence of the disruption, where the abilities or the performance of the system are assessed for each domain (Linkov et al. 2013b). The four phases in which the abilities of the system are assessed are the same as for the original resilience matrix of Linkov et al. (2013a, b):

- Plan/Prepare: phase in which the system has to set the foundation to keep functioning during a disruptive event
- Absorb: phase in which it is necessary to maintain critical functions and service available while isolating the disruption
- Recover: phase in which it is necessary to restore all functions and services to their pre-event functionality
- Adapt: phase where the configuration of the system becomes more resilient using knowledge from the event.

The inclusion of the economic domain is here considered essential, being the necessity of handling the system in its whole the main aim of the proposed methodology. The economic environment is part of the dimension of resilience often addressed in literature (Ribeiro and Gonçalves 2019) and considering its stability it is essential to ensure the overall goal of resilience enhancement (OECD 2021). Disasters or shocks can turn into economic declines both at local level and at higher spatial scales considering current interconnectedness of systems and their sectors where impacts may have cascading impacts. Focusing on economic consequences and perturbations that occur after a stress, or better focusing on the capacity of the economic sphere of a system to prepare, adsorb, recover and adapt allows to monitor properly progress in resilience enhancement over time.

The 16 individual cells of the 4 $\times$ 4 matrix deliver an over-all picture of the functionality of the whole system throughout the phases related to the occurrence of an adverse event. They contain indicators, or more specifically metrics that measure the ability of a system to face the adverse event for assessing its resilience. As specified by Linkov et al (2013b) "the metrics developed through the resilience matrix framework cannot necessarily be measured by direct means and should be generated using specific measures developed on a system-by-system basis. These could be both quantitative data and qualitative measures, which are then evaluated by technical experts and stakeholders associated with the specific system." As explained by Fox-Lent et al. (2015) the resilience matrix "does not express specific metrics, rather it provides a framework to identify the relevant metrics to assess performance from a wider system perspective".

The effort of populating the matrix cell, consists in a diagnosis practice by itself. It allows already to identify gaps during such initial phase, rising attention where the need to allocate resources to enhance resilience lies. Such an approach that envisages a cross- and multi-domain analysis is relatively uncommon; physical elements, functions/services and users/communities are considered in combination. The scores can be developed through different approaches according to Fox-Lent et al. (2015): *Single metric* to handle indicative factors of the performance of the system components, in this case optimal or unacceptable measure for the metric can be used for the performance; *Combined metric* when the functioning is related to multiple features whit different performances, average or weighted sum of multiple metrics can be carried out; *Checklist* approach to develop a score when appropriate indicator of the degree of performance cannot be found; *Expert judgment* in which experts create scores based on their knowledge of the system and experience, scores can be generalized on a scale of none/low/medium/high values for instance or other methodologies can be better developed. By conveying a score to each cell, which represents the ability of the system to handle the circumstances in the specific domain, resilience can be assessed.

In the intersection of raw/column of the proposed revised resilience matrix (Fig. 2.3), each cell is designed to contain a quantitative measures i.e. performance score assigned to each metric. Selected metrics are intended as means for comparing the quality of systems performance or of their components through the application of a measure, where by measure is intended a mean (number or value) to record an attribute of a system or its component (Alberts and Hayes 2005). Therefore, the matrix, through each cell, delivers assigned scores that serve as guidelines to develop objectives for the system under analysis. Metrics, both qualitative and quantitative, intended as means for carrying out the assessment, track and compare the performance of system's processes, must be selected in accordance to each analysed system. They can be defined to assess community resilience, critical infrastructures behaviour or other type of systems of any size, as a urban context. As for the latter, metrics to be assessed could be risk awareness, presence of structural mitigation measures, community participation, data availability and so on.

2.2.2 Stress Testing for Resilience

Stress testing is a method, becoming increasingly widespread, used to analyse the behaviour of a system in case a stressor is applied or a disruption occurs. Stress testing encompasses various techniques for assessing resilience and quantifying vulnerabilities of a system to extreme events, allowing to understand its stability (Čihák 2007). The procedure consists in testing the system beyond standard operational capacity, usually till a breaking point, in order to observe the effects and estimate potential consequences. The stress testing is applied in various disciplines such as finance, medicine, structural engineering, critical infrastructures protection, nuclear science, hardware and software engineering and so on. The concept of stress testing has its roots in medicine, and in the science of material where the test foresees an applied load on materials to determine the resistance characteristics. The nuclear sector makes use of the stress testing; after the Fukushima na-tech accident in 2011 the prerequisite for comprehensive stress testing procedures became an international need. In finance stress testing is part of regulatory framework to assess overall market risk (Pescaroli and Needham-Bennett 2021). For instance, stress testing applied to the bank system allows to analyse how bank institutions would cope with the assumed adverse scenario (i.e. potential future risks), like a recessions or a financial crisis

(Dent et al. 2016). Recently, financial stress testing has been implemented also in connection with climate change impacts (Hallegatte et al. 2022) and also for sectors such as supply chain and computer manufacturing.

A rather wide literature relates to stress testing (Blaschke et al. 2001; Hilbers et al. 2004; Čihák 2004; Čihák and Hermanek 2005). However, fewer studies provide technical guidance on stress tests practical implementation (Čihák 2007), and although the principles of stress testing are simple, its application can be very complex (ECB 2008).

Linkov et al. (2018, 2022) propose combining risk and resilience stress testing in their recent work on critical infrastructures. Our approach focuses on the resilience stress testing, which differs from the more traditional risk stress testing (Mignan 2014) that analyses and quantifies risk factors and it is used as a risk management tool. Stress testing can be used in isolation or in combination with other tools or techniques to enhance situation awareness and improve the understanding of criticalities in a system, while integrating resilience management to allow the system to perform better when a stressor is applied.

Implementing resilience stress testing brings a novel advancement in understanding how systems functionality drops in the aftermath of a disruptive event and how they perform in recovery and adaptation phases. The stress testing applies a stress (a disruption) on the system and through the variation of the amount of such disruption it is possible to understand how the systems responds to it.

2.2.3 Applying the Stress Testing to the Resilience Matrix for Complex Urban Systems

Considering that some systems are by nature complex and characterized by multifaced interconnections, even a small amount of stress may significantly impact their functioning. An approach that does not focus only on one or few components of the system would fail in assessing its resilience. Resilience is function of the interaction across multiple levels of functioning and the lack of resilience at one level might weaken resilience at other levels (Alexander 2013). The stress experienced by one functional sub-area of the system might require more resources than in normal conditions to keep functioning, making use of the means of the whole system. Thus it might create a condition in which due to the induced stress on the other sub-areas, the systems saturates and collapses. Stress testing depicts the point at which the system is not able to handle the amount of load imposed on it so its performance drops down rather abruptly. Resilience addresses the mode in which the system reaches a new level of functioning which might be the same as before, lower or in some favourable cases higher, therefore, how the system is able to cope with the applied stress.

Accordingly, considering stress tests as useful tools to assess resilience of systems, the resilience matrix becomes a complementary operational tool that allows to quantify such resilience in all its stages by handling the system in its whole. The matrix makes possible the integration of the stress testing results within its structure to evaluate the system behaviour in terms of preparedness, in the initial phase of the disruption (adsorption) and then assess the subsequent phases (recovery and adaptation). This comprehensive approach makes possible to identify where the criticalities of the system lie (domains) and when they arise (phases) due to the stress applied. Linkov et al. (2022) propose the resilience matrix as a semi-quantitative modelling tool to identify interdependencies in a system which must be used in conjunction with other analytical tools to carry out a stress testing for critical infrastructures.

The combination of the two tools is proposed as an innovative methodology to assess the resilience of a system, such as the urban system, with the aim to support spatial planning.

Stress testing needs to be adaptable to be able to cover the complexities of handled systems. Systems vary in their structure and dynamics, therefore their characterization varies too. Most importantly, as proved by this work, stress testing has to effectively cope both with lack of data and with the uncertainty of available ones. The methodology proposed is well structured but flexible to combine quantitative and qualitative data. The resilience matrix allows to consider the temporality of resilience, examining how the system behaves after a stress has been applied alongside the domains that compose a complex system. Through the resilience matrix suitable results can be obtained given its nature and methodological features.

2.3 Implementation of the Proposed Methodology

The proposed methodology, which makes use of the stress testing and organizes its results in the resilience matrix, consists in a procedure to operationalize the evaluation of urban resilience in areas exposed to hazardous events making use of qualitative and quantitative data together with spatial and non-spatial ones with standardized steps. Indicators are used to assess through a semi-quantitative process the resilience factors considering not only physical, but also social and institutional components of the urban system.

The procedure based on data investigated at different spatial scales, from regional (Valencian Community) to local (municipality of Benidorm and postal code section), has been applied to the case study of Benidorm.

By stressing the system under consideration, resilience is assessed and framed within the resilience matrix. The latter is a tool for strategic handling of data and information, which allows to increase the degree of systematisation of the results and to deliver the results in a visual fashion. It consists in a deep analysis of the urban system to identify critical areas on which it is necessary to act in order to increase resilience through different phases of disaster risk management and to highlight what capacity is available to the system to respond/withstand and react in case of disruptions.

The methodology has been implemented and tested through three different scenarios (Stress testing A, B, C) described in detail in Chap. 4. Once defined the system, precisely the urban area of Benidorm in this application, three different stressors have been applied on its functioning. Specifically they can be considered as three scenarios associated to present, past and future. The scenario associated to the present stresses the system under consideration, in its current configuration, examining and questioning risk data (i.e. hazard, exposure and vulnerability data) availability and effectiveness. The scenario associated to the past stresses the system by applying a past flooding event occurred in 2007 and making consideration about the damages which were repeated following very similar dynamics 7 years later. While the last scenario stresses the system by considering the potential future impact of climate change on the current situation of the system. For the three stresses the behaviour of the system is evaluated through a score and then organized into the resilience matrix which allows to show visually the performance of the system for each domain of a complex system and in each of the prepare, adsorb, recover and adapt phases.

2.3.1 *The Operational Steps*

For each stress testing, the procedure followed is characterized by the same main steps.

The procedure can be applied to different areas by customising the elements of the system's domains with proper inputs data and scenarios. Such methodology would facilitate the integration of resilience management within spatial planning operations.

The replicable operational steps followed to assess the resilience of the system through stress testing and resilience matrix be recapitulated as follow:

1. DEFINITION OF THE SYSTEM—This step foresees the characterization of the system analysed: definition of spatial boundaries, identification of critical functions and critical system's components that are essential to the system operation, evaluation and assessment of present conditions. When available risk, exposure, hazard and vulnerability (physical, social, systemic) can be used as inputs data. According to the specific case of analysis different input figures can be used ranging from maps, charts, results of interviews and surveys, statistics, reports and other significant data available. An evaluation of factors and parameters involved and their interrelations needs to be carried out to understand how the system functions in ordinary times. This step requires to identify the proper input information about the system to have a comprehensive picture of the status of the system. Urban resilience analysis requires a comprehensive understanding of the complex system of the city (Aguilar-Barajas et al. 2019).
2. DEFINITION OF THE STRESSOR—This step requires to identify the stress that is going to be applied to system in order to test it. The assumptions for the definition of the threat derive from the knowledge of the system and available data.
3. APPLICATION OF THE STRESSOR AND EVALUATION OF THE EFFECTS THROUGH SELECTED METRICS—This step consists in the operational

execution of the stress-testing. A simulation of the application of the stressor is carried out to evaluate both positive (eustress) or negative (distress) effects. Once the stress is applied its effects are assessed on each metric to understand and uncover how the system performance is impacted by the disruption, by analysing the conditions of the critical components and their interrelations conditions. At this step also the possibility of cascading, enchained impacts is considered. Metrics are selected to track and determine along the stress testing process if the system in its whole (i.e. combination of different critical functions belonging to different domains of a complex system) meets resilient performance goals (see step 4). More resilient systems are those that are able to diminish the disruption potential of cascading hazards and multiple enchained impacts.

4. RESILIENCE ASSESSMENT. ASSESSMENT OF THE PERFORMANCE AFTER THE DISRUPTION WITHIN THE RESILIENCE MATRIX—In this step the resilience of the system is assessed by evaluating its performance after the disruption has been applied. This phase requires to generate scores of the configuration and its functioning. This step consists in the modelling of the post-disruption process where temporal indicators of how the system responds to the disruption are set. The parametrized results are systematized in the cells of the matrix per each domain according to the disaster phase to narrow down the cause of the performance problem, and to find an appropriate solutions.

Once the process is completed, the obtained results allow to identify where improvements are needed. Such methodology can be considered as an inspection tool to evaluate the system behaviour in case of disruption. Stress testing through resilience analysis matrices permits to understand where and at what disaster phase problems arise and how an impact on one function can be contained not to influence others, to assist in the development of enhanced risk and resilience management plans.

Figure 2.4 conceptually illustrates the proposed methodology which sees three main sections (i) input data; (ii) simulation; (iii) results. Essentially the first section regards the first two operational steps (definition of the system and of the threat/stressor); in the second section the simulation of the resilience stress testing takes place i.e. the fourth operational step of application of the stressor and evaluation of the effects; and, in the last section the resilience assessment takes place, where operationally the performance of the system after the disruption is understood once the results are schematized and organized within the resilience matrix.

Figure 2.5 illustrates how the proposed methodology has been implemented. The table contains in the first column the parameters to be evaluated, i.e. the selected metrics identified to assess the performance of critical functions with a system perspective, once the operational steps 1 and 2 have been completed. Then, the identified stress/threat is applied on each parameter which represents a critical function/characteristic and a score is assigned to each of them in function of the evaluated behaviour/performance of the system for the specific metric present in each raw. The score, according to the procedure designed, varies on a scale from 1 to 4 (second main column), where each integral number within the range means—very low, low,

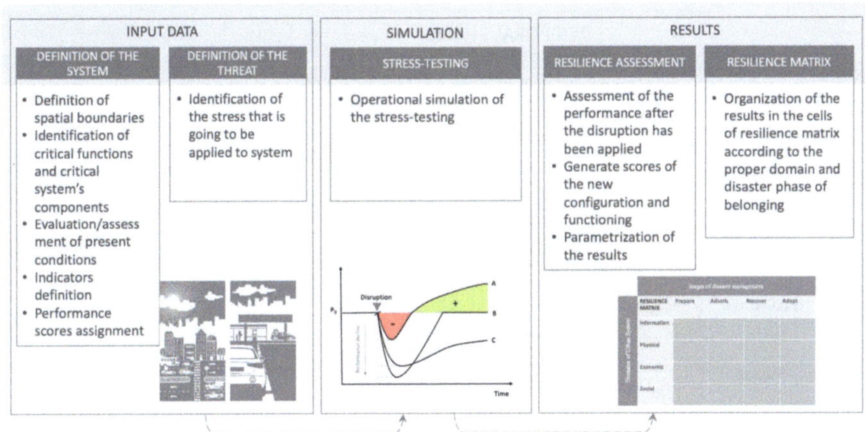

Fig. 2.4 Operational steps of the methodology implementation

moderate and high—performance level. The score is here assigned in function of the understanding of the system. It is in this phase that the simulation of the stress testing (which makes use of the input data) happens; the selected metrics represent critical factors of the system and the score assigned represent the behaviour of the system when the stress is applied.

Then for each row the position of allocation within the resilience matrix is specified (third column), when there is no allocation in one of the 16 cells due to the availability of the parameters considered then the cell is left white, meaning not available performance score since no metric has been identified for the exact intersection domain/ phase. Following the procedure allows to collect additional comments, if any, and the source of the information of the metric from which the performance score has been retrieved (fourth and fifth columns). If more than one performance score is allocated

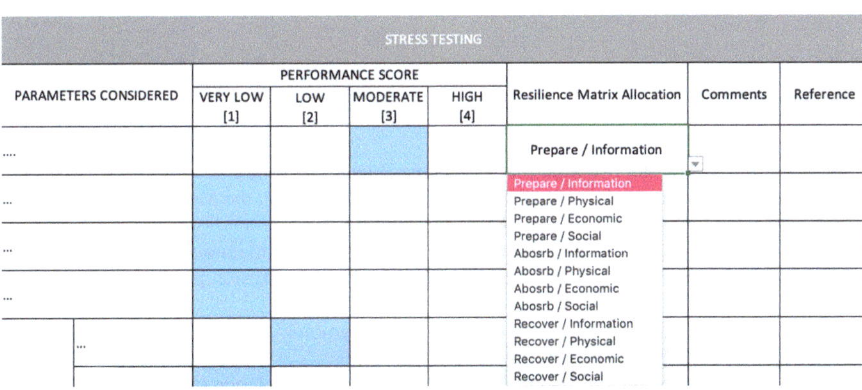

Fig. 2.5 Implementation of the methodology (simulation of the stress testing and allocation of the resilience assessment for each metric in one of the cells of the resilience matrix)

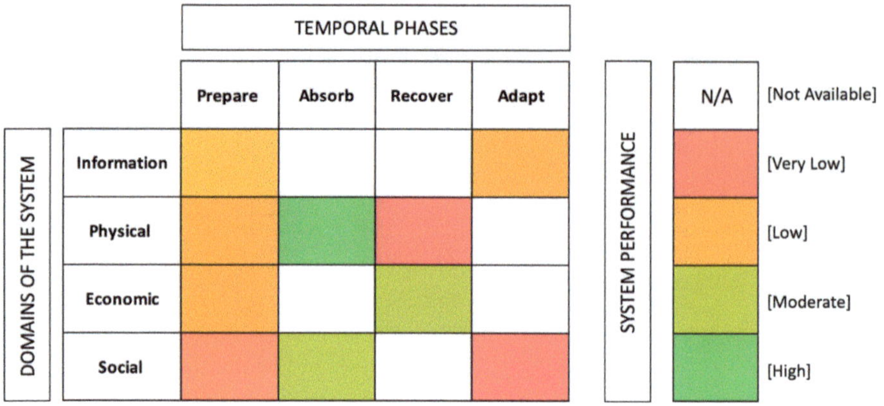

Fig. 2.6 Sample of the final resilience matrix

in the same cell of the matrix, they are joint with an average, i.e. sum of the performance scores allocated in the same cell specified in the Resilience Matrix Allocation column ($a_1 + a_2 + \ldots + a_n$) divided by the number of parameter considered (n), and the final resilience matrix is built, through a pre-set excel sheet, as the one in Fig. 2.6 which represents a sample of a possible outcome of the procedure. In this phase the organization of the assessment of the performance of the system after the disruption/stress is systematized within the resilience matrix.

Each cell allows to understand the ability of a system to manage adverse events, addressing the question: "How is the system's ability to [plan/prepare for, absorb, recover from, adapt to] a disruption in the [information, physical, economic, social] domain?". The column of plan/prepare regards the understanding and documenting of the system, in terms of its structure and functions; while the columns of absorb/recover/adapt depict the abilities that a system should possess in order to be resilient in case of an adverse event (stress).

References

I. Aguilar-Barajas, N.P. Sisto, A.I. Ramirez, V. Magaña-Rueda, Building urban resilience and knowledge co-production in the face of weather hazards: flash floods in the Monterrey Metropolitan Area (Mexico). Environ. Sci. Policy **99**, 37–47 (2019)

D.S. Alberts, R.E. Hayes, Campaigns of experimentation: Pathways to innovation and transformation, in Assistant Secretary of Defense (c3i/command control research program) (Washington DC, 2005)

D.E. Alexander, Resilience and disaster risk reduction: an etymological journey. Nat. Hazard. Earth Syst. Sci. **13**(11), 2707–2716 (2013)

P.M. Allen, Cities and regions as evolutionary, complex systems. Geographic. Syst. **4**, 103–130 (1997)

L.M. Bettencourt, Cities as complex systems, in *Modeling Complex Systems for Public Policies* (2015), pp. 217–236

W. Blaschke, M.T. Jones, G. Majnoni, M.S. Martinez Peria, in *Stress Testing of Financial Systems: An Overview of Issues, Methodologies, and FSAP Experiences* (2001)

M. Čihák, Stress testing: a review of key concepts, in *Research and Policy Notes* (2004) (2004/02)

M. Čihák, *Introduction to Applied Stress Testing*. IMF Working Paper No. 07/59 (2004). Available at: https://ssrn.com/abstract=973989

M. Čihák, J. Hermanek, *Stress Testing the Czech Banking System: Where are We? Where are We Going*? (No. 2005/02). Czech National Bank (2005)

K. Dent, B. Westwood, M. Segoviano Basurto, in *Stress Testing of Banks: An Introduction* (Bank of England Quarterly Bulletin, Q3, 2016)

ECB, in *European Central Bank, Financial Stability Review* (2008). ISSN 1830–2025. Available at: https://www.ecb.europa.eu/pub/pdf/fsr/financialstabilityreview200812en.pdf

D.A. Eisenberg, I. Linkov, J. Park, M.E. Bates, C. Fox-Lent, T.P. Seager, Resilience metrics: lessons from military doctrines. Solutions **5**(5), 76–87 (2014)

C. Fox-Lent, I. Linkov, Resilience matrix for comprehensive urban resilience planning. Resilien. Orient. Urban Plann. Theoretic. Empiric. Insights 29–47 (2018)

C. Fox-Lent, M.E. Bates, I. Linkov, A matrix approach to community resilience assessment: an illustrative case at Rockaway Peninsula. Environ. Syst. Decis. **35**(2), 209–218 (2015)

S. Hallegatte, F. Lipinsky, P. Morales, H. Oura, N. Ranger, M. Gert Jan Regelink, H.J. Reinders, *Bank Stress Testing of Physical Risks under Climate Change Macro Scenarios: Typhoon Risks to the Philippines* (2022)

M.P.L.C. Hilbers, M.M.T. Jones, M.G.L. Slack, in *Stress Testing Financial Systems: What to do When the Governor Calls. International Monetary Fund* (2004)

D. Koren, V. Kilar, K. Rus, Proposal for holistic assessment of urban system resilience to natural disasters. in *IOP Conference Series: Materials Science and Engineering* (Vol. 245, No. 6, IOP Publishing, 2017), p. 062011

B.D. Linkov, J. Trump, G. Trump, W. Pescaroli, A. Hynes, A.P. Mavrodieva, Resilience stress testing for critical infrastructure. Int. J. Disast. Risk Reduct. (2022). https://doi.org/10.1016/j.ijdrr.2022.103323

I. Linkov, D.A. Eisenberg, M.E. Bates, D. Chang, M. Convertino, J.H. Allen ... T.P. Seager, in *Measurable Resilience for Actionable Policy* (2013a)

I. Linkov, D.A. Eisenberg, K. Plourde, T.P. Seager, J. Allen, A. Kott, Resilience metrics for cyber systems. Environ. Syst. Decis. **33**(4), 471–476 (2013b)

I. Linkov, C. Fox-Lent, L. Read, C. Allen, J.C. Arnott, E. Bellini, ... D. Woods, Tiered approach to resilience assessment. Risk Anal. **38**(9), 1772–1780 (2018)

I. Linkov, B.D. Trump, W. Hynes, Resilience strategies and approaches to contain systemic threats, in *Conference paper* (2019

S. Menoni, F. Atun, in *Cities and DRR. In: Educen Culture & Urban Disaster. A Handbook. The Netherland* (2017). ISBN:978–94–6343–600–7

A. Mignan, MATRIX Common IT sYstem (MATRIX CITY) Generic multi-hazard and multi-risk framework-the concept of Virtual City-IT considerations. New Multi-Haz. Multi-Risk Assess. Meth. Europe (2014)

National Academy of Sciences, *Disaster resilience: a national imperative* (United States, Washington DC, 2012)

OECD, *Fostering Economic Resilience in a World of Open and Integrated Markets Risks, Vulnerabilities and Areas for Policy Action*. Report, prepared by the OECD for the UK Presidency of the G7 (2021). Available at https://www.oecd.org/newsroom/OECD-G7-Report-Fostering-Economic-Resilience-in-a-World-of-Open-and-Integrated-Markets.pdf

G. Pesaro, M.T. Mendoza, G. Minucci, S. Menoni, Cost-benefit analysis for non-structural flood risk mitigation measures: Insights and lessons learnt from a real case study, in *Safety and Reliability–Safe Societies in a Changing World* (CRC Press, 2018), pp. 109–118

G. Pescaroli, C. Needham-Bennett, Operational resilience and stress testing: Hit or myth? J. Financ. Transform. **53**, 32–43 (2021)

C. Pursiainen, B. Rød, G. Baker, D. Honfi, D. Lange, Critical infrastructure resilience index, in *Proceedings of the 26th European Safety and Reliability Conference, Glasgow, UK* (Vol. 26, 2016), pp. 2183–2190
P.J.G. Ribeiro, L.A.P.J. Gonçalves, Urban resilience: a conceptual framework. Sustain. Cities Soc. **50**, 101625 (2019)
P.E. Roege, Z.A. Collier, J. Mancillas, J.A. McDonagh, I. Linkov, Metrics for energy resilience. Energy Policy **72**, 249–256 (2014)
T.L. Root, S.H. Schneider, Ecology and climate: research strategies and implications. Science **269**(5222), 334–341 (1995)
T.J. Wilbanks, R.W. Kates, Global change in local places: how scale matters. Clim. Change **43**(3), 601–628 (1999)

Chapter 3
The Case Study

Abstract This chapter focuses on the analysis of the peculiarities of the case study selected. The city of Benidorm is introduced, followed by a review of the urban planning progress and an outline of the hazards and risks affecting the area of interest. The city is exposed to different hazards such as forest fire, landslide, earthquake but most importantly flash floods. Its exceptional urban growth, population increase and extraordinary number of inhabitants/day during the high season characterize it as a model of high exposure and vulnerability. The chapter concludes with a detailed discussion of the data collection and processing approaches.

3.1 The City of Benidorm: A Spanish Fast-Growing (Mass) Tourism-Specialized Coastal City in Hazard Prone Area

While thinking about a node where interactions among nature, society and its functions and assets, as well as economic activities are embodied, the case study of Benidorm, a mass tourism-specialized coastal city in hazard prone area has been selected, being an emblematic model where observations, analysis and implementations could have been carried out due to its distinct characteristics. The city is exposed to different hazards such as forest fire, landslide, earthquake but most importantly flash floods. Its exceptional urban growth, population increase and extraordinary number of inhabitants/day during the high season has led to an increase of both exposure and vulnerability to the hazards threatening the area. Direct and in-loco interactions with the researchers and the stakeholders of the area, and site-visits took place. As a result, a deep analysis of the area allowed to understand the impact of past natural and socio-political events on the current configuration that served as a basis to implement and test the developed methodology. Three different stress testing have been applied to the system and the results organized in three respective resilience matrixes to highlight areas where criticalities lie and efforts are required to boost resilience and address efficient actions aimed at reduce disaster risk.

A. Faiella, *Learning from Disasters to Enhance Resilience of Urban Complex Systems*, PoliMI SpringerBriefs, https://doi.org/10.1007/978-3-032-07356-3_3

Multiple data sources have been used for the analysis of the case study, performing direct interactions with researchers who have conducted analysis on the area of interest, local stakeholders and in field observation plaid a fundamental double role in the whole research path. As matter of fact, the construction of the case study is not only an object of the research but also and most importantly an instrument through which the proposed methodology has been developed and shaped, then applied and tested (see Fig. 2.2).

The city of Benidorm is located on the Spanish east Mediterranean coast, 40 km north of Alicante and 120 km south of Valencia. A city of 3550 hectares, with 12.3 km of coastline, of which 5.3 are beaches (Mazón et al. 2012), located in the southern part of the Valencian Community, in the Alicante Province. The Valencian Community (Comunitat Valenciana) is an autonomous community of Spain and is the fourth most populous Spanish autonomous community, located alongside the Mediterranean coast on the east side of the Iberian Peninsula and divided in three provinces which are Castellón, Valencia and Alicante. Given its climatic and atmospheric conditions and geographic characteristics, the territory of the Valencian Community is prone to different natural hazards such as seismic activity, landslides, heavy and torrential rains alternating with lack of precipitations, extreme temperature variations, droughts, winds and hailstorms. In addition to the physical environment's characteristics, the Valencian Community has experienced a growth in terms of exposure, considering the occurred exceptional anthropic development, in terms of urban expansion and infrastructure construction facilitated by the different varying regulations in force in terms of land use and urban planning. The occupation of fluvial-plains and waterways for agricultural, urban and touristic activities have increased the vulnerability of the territory. Modern agricultural practices which abandoned the percolation of flood water to be re-distributed rapidly and homogeneously to agricultural fields has increased the amount of solids carried by the waterways and the intensification of erosion on slopes. Forest fires that often occur in the region leave burned hundreds hectares provoking soil erosion and increase of sediment transport by floods (Olcina Cantos 2010). Considering the risk as a function of hazard, vulnerability and exposure, the Valencian Community is one of the most risky regions in the whole country (Olcina Cantos 2010). Records show that both direct and indirect damages, in terms of physical, environmental and human impacts, have increased significantly from the 80's to the present (Palencia Jiménez et al. 2018).

Benidorm is a mass-tourism summer destination, one of the most known mass tourism target on the Spanish Mediterranean coast (see Fig. 3.1). From the mid-1950s, the city has invested heavily in the tourism industry: the urban master plan was actually designed to enhance the tourism business (Nolasco-Cirugeda et al. 2020). Nonetheless, Benidorm is considered by some scholars as a good practice of sustainable urban development to support the mass-tourism industry. Our view though is that Benidorm is a case that deserves deeper analysis not only because of its striking urban image, but also for the intense urban expansion that caused an increase of exposed asset and people to natural hazards (Martínez-Ibarra 2017) and

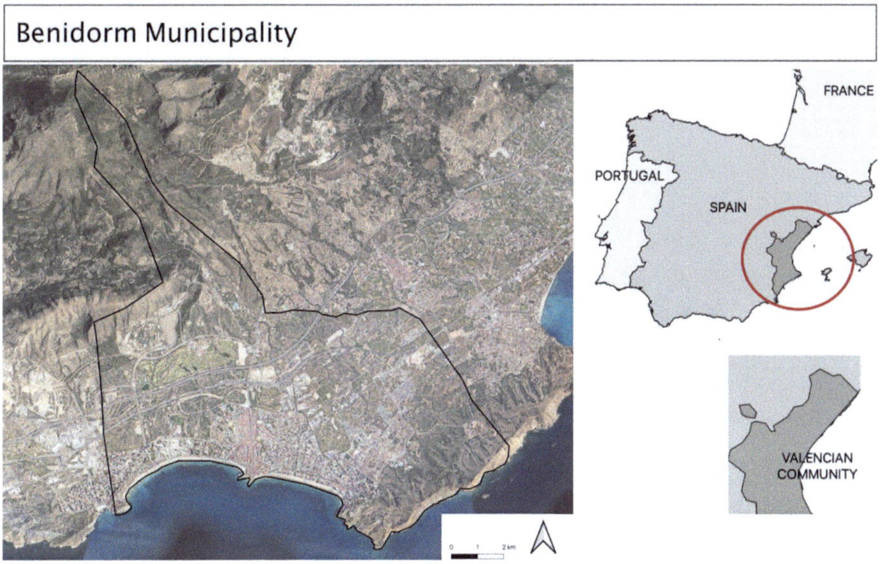

Fig. 3.1 Geographical location of Benidorm

for the management of urban runoff which often occurs after intense rainfalls and water resources in general which represent one of the most significant challenges for cities located in the Mediterranean coast especially in the face of climate change effects (Sánchez-Almodóvar et al. 2023; Olcina Cantos and Hernández Hernández 2021).

Before the 1950s, Benidorm was a fishermen village. In 1950 national tourism started in the village and six years after the city issued its first General Urban Development Plan establishing the growth of the city around the main beaches. First hotels opened and in 1959 the Mayor of Benidorm allowed the use of bikini in the beaches of the town. In 1963 Benidorm issued its second General Urban Development Plan which allowed higher skyscrapers. In 1970 the airport of Alicante was opened and five years later the motorway between Alicante and Benidorm was built. Nowadays according to the Municipal Territorial Plan against Emergencies of Benidorm (MTP 2019) the normal population is around 66,831 reaching as many as 215,000 inhabitants/day during the summer season. The statistics, such as also the following Figs. 3.2 and 3.3, reflect the intense transformation of the coastline of the province of Alicante that has occurred since the 60s of the twentieth century.

The main urban settlements developed alongside two beaches, respectively Levante and Ponente beaches (see Fig. 3.4); an area where the weather is mild and most of the year being not affected by easterly and cold northern winds.

High-rise buildings distinguish Benidorm's image, which was named "the most effective mass tourism machine in Spain" (Maas et al. 2000) with the highest rate of skyscrapers per capita in Spain (Nolasco-Cirugeda et al. 2020). Born as a site for elite, changes over time have allowed different working classes, also with diverse

Fig. 3.2 Benidorm before becoming a mass-tourism destination (*Source* Town Council of Benidorm[1])

economic profiles to benefit from the touristic offers. Several second homes have been built too (Prieto-Arranz and Casey 2014; Arnstberg and Ergstrom 2007; Domínguez et al. 2016). Benidorm is the third European city in terms of skyscraper concentration, after London and Milan, with a total of 310 buildings over 35 m high (MTP 2019).

3.2 Developments of Benidorm's Urban Planning

Spain is a state constituted by 4 levels of government: (i) the national government, (ii) autonomous communities, (iii) provinces and (iv) municipalities. The constitution specifies the authority regarding land-use policy assigning responsibility for spatial planning to the autonomous communities, although, national government prepares framework legislation that guides regional laws, sectoral plans for national infrastructure and addresses environmental legislation and related legislation that affects the possibilities to develop land but it has it has no authority to prepare a general national spatial plan. Autonomous communities enhance the national framework legislation by establishing individual legislative framework on land-use planning. Regional governments prepare regional spatial plans to which municipal governments are tied, however, municipalities are the main actors in land-use planning.

[1] https://rua.ua.es/dspace/bitstream/10045/27662/6/benidorm_evolution.pdf.

Fig. 3.3 Benidorm nowadays (*Source* Town council of Benidorm)

The Spanish urban planning policies resemble the political changes occurred over the years. Since the first Land Use Act (1956), legislation and frameworks have been revised and re-adapted according to the changing socio-economic, political-administrative and legal context. The year 1978 marks a shift from central government regulating urban planning to a decentralization of powers distributed between regions and municipalities; a change mainly attributed to the Spanish constitution of 1978, alongside the call of the European Union from the mid-1980s towards a bottom-up land development and the following Local Government Standards Act of 1985. This, made urban planning a key subject in the local politics of the recent democratic municipal councils (Pérez 2007).

Most relevant changes of deregulation occurred between the end of the second stage and in the last one were a consequence of the Law 6/1998 (Bellet Sanfeliu 2021; Pérez 2007). Law 6/1998, as stated by Bellet Sanfeliu, became informally known as the "law that allowed everything to be developed" since, instead of regulating urban growth, it allowed potential urban development in every land that was not clearly protected encouraging urban production. Following, the National Land Law 8/2007 consolidated the Land Use and Urban Planning Act but with insufficient dominance considering what occurred in the previous years. Likewise, Fariña and Naredo (2010) highlight that the history of contemporary Spanish urban planning grounds on the Land Act of 20 June 2008. They state that the construction sector consists in the principal driving force of the Spanish economy, a consequence the years between 1998 and 2007, in which Spain experienced an extensive development in terms of buildings due to the deregulating legislative choices which encouraged

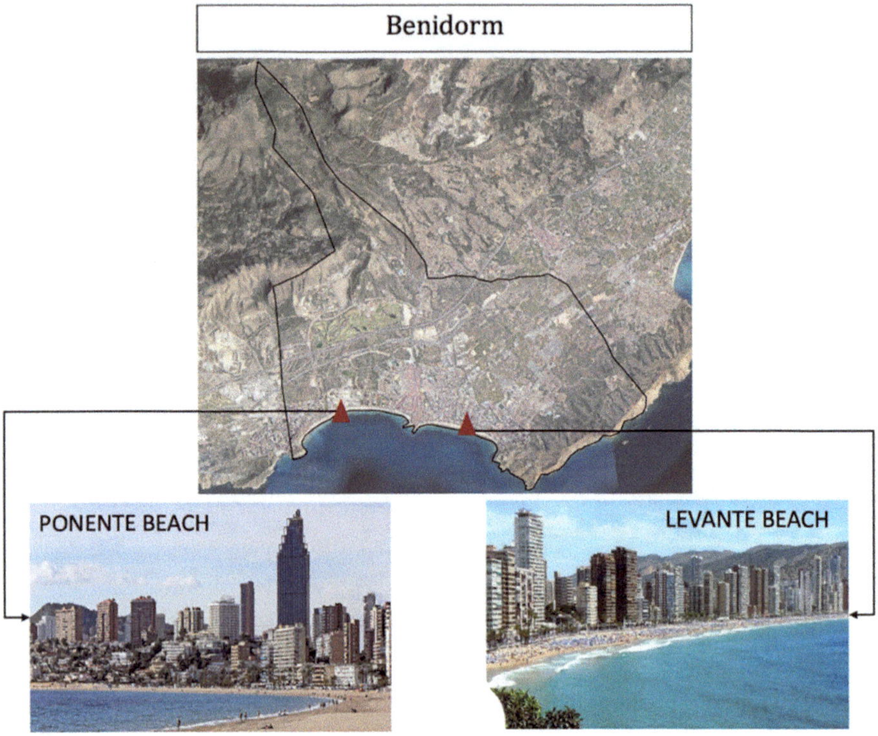

Fig. 3.4 Benidorm beaches (*Source* of the pictures https://en.visitbenidorm.es)

widespread development across the country. The Land act of 2007 stepped forward a more sustainable approach, addressing the need to adopt a new model focused on environmental concepts related to the value of rural land. The current regulation in force in Spain is the Royal Legislative Decree 7/2015 of October 30, which approves the consolidated text of the Land Law and urban rehabilitation.

It is in this context that the developments of urban planning in Benidorm got shaped.

The General Plan of Benidorm was one of the first delineated and approved urban plan in Spain after the 1956 Land Act came into force. The first masterplan (1956) aimed at developing a "new city" characterized by grid pattern layout inspired to Howard's garden city model (1898). A garden city model has been thought by Ebenezer Howard as a city with proper dimensions to allow development of social life; surrounded by greenbelts, of public property or managed on behalf of the society, such model however has been also criticized for damaging the economy, being destructive of the natural environment identity (Edwards 1914). Benidorm new shape focused on low building density, to be developed alongside the two beaches, Levante and Poniente, offering dynamic public spaces, wide road axes, car-parking and avenues rich in lined trees (Ciriquián and Meyer 2004), as wells as safeguarding the initial urban area for the local population and the inland area to preserve its agricultural value (Vera 1987). The first Masterplan of 1956 has varied and evolved along the years since its approval (Rodríguez 2003). Around the fifties internal tourism in Spain incentivized the establishment tourist settlement in Benidorm along the Levante beach area (Zaragoza 1955). Developers and landowners, once understood the potential of the tourism business, persuaded the local administration to remove the restrictions on building height; modifications to the 1956 masterplan related to building height were approved in 1963. The model, which distinguish the current image of Benidorm, was based on Le Corbusier's proposal in Manhattan, New York (Le Corbusier 1937); where skyscraper should be "a function of capacity and of the area of free ground at its base". Between the 1970s and the 1980s a regeneration of the image of the city as a tourist destination has been promoted handing urban revitalization and public spaces over to private companies and their marked-based principles (Fraguell et al. 2016), named by Nolasco-Cirugeda et al. (2020) "the Rousification of Benidorm" subsequent Jim Rouse's concepts and the "the Rousification of America" chapter in Cities of Tomorrow written by Hall (2014).

Therefore, nowadays, most emblematic characteristic of the urban development of Benidorm, to support the development of the tourist industry, resides in the vertical growth. High-rise buildings distinguishing Benidorm's masterplan (Nolasco-Cirugeda et al. 2020).

Until 1960 Benidorm, a small fishing village, recorded 6000 inhabitants. In 2020, according to the census, the permanent population of Benidorm reached 70,450 inhabitants. This value, does not include the fluctuations of visitors, which are concentrated especially in summer when the total population of the city can reach up to 400,000. The first bungalows were built on the Levante beach, in 1925, but it is in the fifties that the real transformation of the space has started through a model of exploitation of resources (Del Castillo 2014).

The new masterplan of 1990 (Plan General de Municipal de Ordenación Urbana, PGMO[2]) was the instrument needed to increase tourist quality standards (Baidal et al. 2013). Nonetheless, within different measures and regulations, the new plan facilitated private developers in investing in new leisure proposals. It was in this period, according to Nolasco-Cirugeda et al. (2020) that controversial decisions were made as a consequence of the forest fire that devastated large land surface in Sierra Cortina. Public investments allowed the construction of theme parks, golf fields and other leisure attractions and modifications to the 1990 masterplan approved moved against the compact urban model implemented in the initial 1956 masterplan. On August 11, 1992 a forest fire destroyed the pine forest of the Sierra Cortina, known as "the largest pine forest in the Mediterranean" an area that different administrations of the Marina Baixa area tried to preserve by opposing the urbanization of what was considered "the green lung of Benidorm". This was a rough period, according to Soto (2017) for Spain where requalification, especially on the coast, has been strongly

[2] https://benidorm.org/es/ayuntamiento/concejalias/urbanismo/arquitectura/plan-general/plan-general-de-municipal-de-ordenacion-urbana-pgmo/plano-no3-regimen-proteccion-Y-gestion-del-suelo.

influenced by questionable political actions. Different fires occurred and the origins were not investigated or the causes were not determined. Three years after the fire, in 1995, the Terra Mítica project designed to be developed in that area was declared of interest to the Valencian Community and a private company in 1996 started a project of a huge hotel zone, two golf courses and a theme park on the land devastated in the fire. The Terra Mítica project, which made also use of public money, became private property, both in terms of land and facilities. In 2003, the Forestry Law prohibited the requalification of burned land for 30 years. According to the Fundación Civio, a small amount of fires that occurred between 2001 and 2013 were caused to obtain a change in land use (only 0.15%).

Benidorm is nowadays a city with the highest rate of skyscrapers per capita in Spain, after Madrid and Barcelona.[3] The General Plan for Urban Planning focused on the coordination of both urban regulations establishment and tourism expansion. Instead of designing a low-density developments model, the regulations endorsed the erection of high-blocks reserving, around the towers and on some stretches of the coast, open spaces (Mazón et al. 2012). Such model, is considered by some as an efficient land occupation strategy (Ciriquián et al. 2018). The urban planning regulations, designed in response to the tourism sector demand allowed, to make Benidorm a successful tourist center (Mazón et al. 2012).

Benidorm urban development has received different feedbacks, varying from scholars which considered Benidorm as a remarkable urban phenomenon (Lefebvre 2014; Mazón et al. 2012; Maas et al. 2000, Nolasco-Cirugeda et al. 2020) and others who criticized its features, such as contrast between the skyscrapers, massive buildings evident from far distances, and the natural landscapes of seaside and almost desert arid areas typical of the area. In addition to the critiques related to the seaside landscape which would have been better preserved by building a less striking urban profile (Fernández-Galiano 2002), also for the intense urban expansion that caused an increase of exposed asset and people to natural hazards remains a significant issue to be addressed (Martínez-Ibarra 2017).

As mentioned, the original strategy, was to occupy only the coastal strip with an intense and compact urban development; subsequently, following the rapid growth of Benidorm, in order to answer to the needs of the touristic demand, a general occupation of the inland municipal area took place (see Fig. 3.5). The variation of the land use criteria in the 1990s resulted therefore in less sustainable developments inland due to the occupancy of wide areas dedicated to leisure attractions and accommodation infrastructures through a model, this time, more similar to the neighbouring municipalities.

[3] https://skyscraperpage.com/database/country/13.

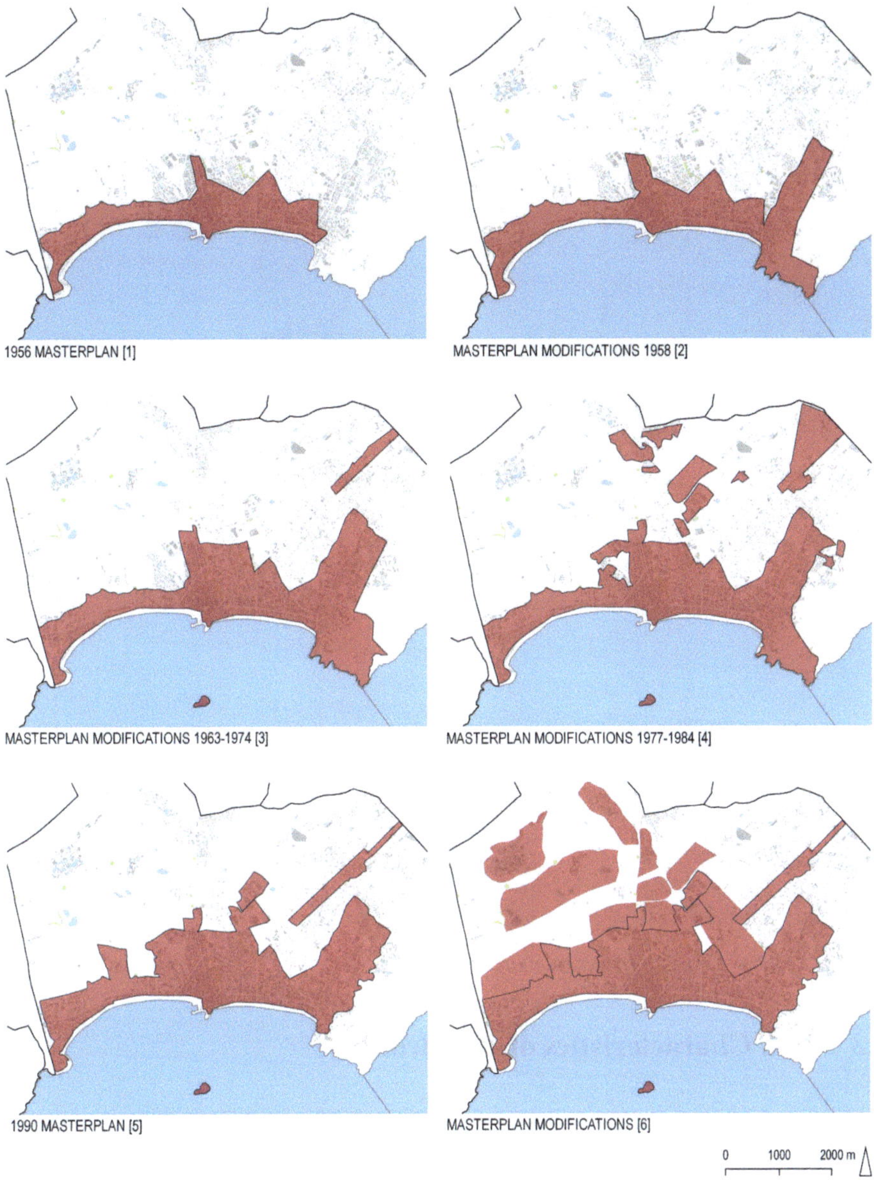

Fig. 3.5 Land occupation in function of the modifications to the masterplan (*Source* Nolasco-Cirugeda et al. 2020)

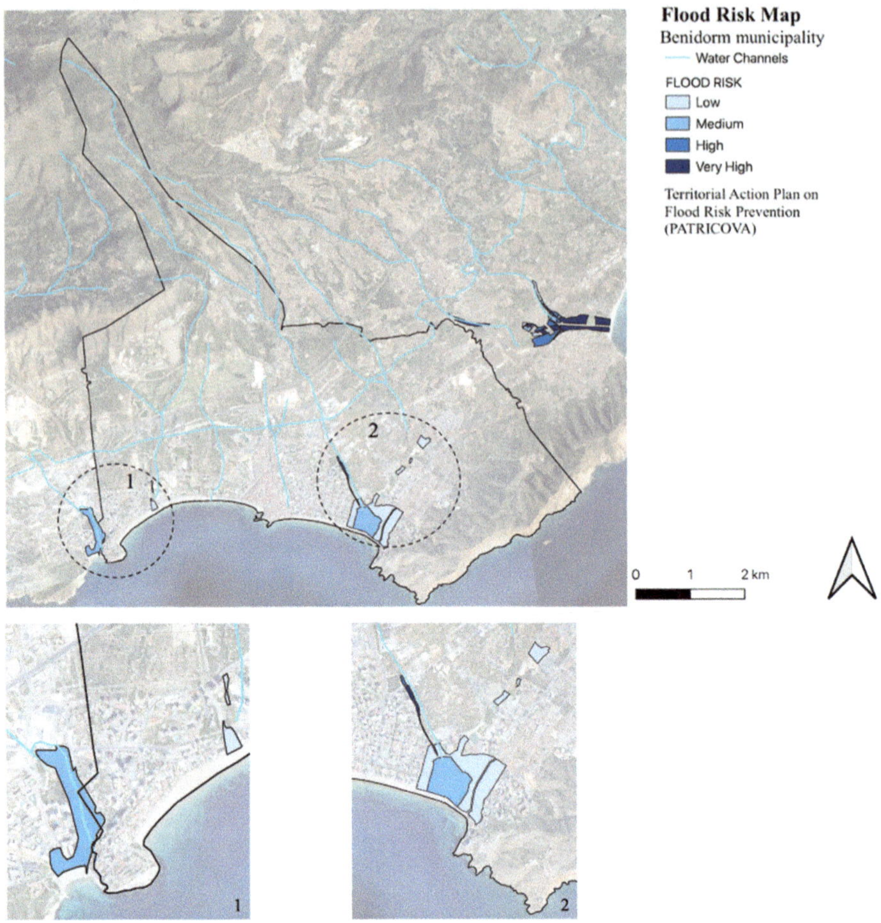

Fig. 3.6 Flood risk map of Benidorm municipality

3.3 Risk Characteristics of Benidrom

According to MTP (2019) and in line with the characteristics of the Valencian Community area where Benidorm is located, the city is exposed to the following hazards:

- Forest Fires: urbanized areas close to the forest masses. Benidorm city has a total surface of 3.928,92 hectares, of which 1.780 are forest area (Plan especial riesgo incendios 2010). Following the MTP (2019), the main forests are those of the Natural Park of Sierra Helada, Sierra Cortina, Partida Murtal, Partida La Tapia and Partida Foia del Ca. Those areas are categorized in the Special Plan against the Risk of Forest Fires of the Valencian Community with an Extreme Relative

Risk and as a High Risk Zone. Recent most significant forest fires have occurred in 1992 and 2009;
- Droughts: in 1978 a severe drought has been suffered, water was supplied by means of tanker boats that, anchored in the bay of Benidorm, were discharging the water into the municipal tanks through pipes appositely installed;
- Landslides: in the municipality of Benidorm there are areas of low and medium risk of landslides;
- Floods: there are several areas in the municipality where floods usually occur. The most important flood that the municipality of Benidorm has suffered occurred in October 1971 with significant human and material losses. Other major floods occurred in 1983, 1984, 1985, 1986, 1987, 1989, 1990, 1996, 1997, 1998, 2002, 2007, 2008, 2010, 2012, 2013, 2015, 2017. According to the Special Plan against the risk of flooding of the Valencian Community (El Plan Especial frente al riesgo de inundaciones de la Comunitat Valenciana), the risk of the municipality is classified as "*medium*". The most important flood that Benidorm has suffered with significant human and economic losses occurred in October 1971. In 1990 a Basic Municipal Plan in force against the Risk of Fooding (Plan Básico Municipal vigente frente a Riesgo de inundaciones) has been elaborated, this led to various actions aimed at improving the drainage in some areas of the city, although there are still several areas in the municipality where floods usually occur (MTP 2019). More recently, the Territorial Action Plan on Flood Risk Prevention of the Valencian Community (PATRICOVA[4]), has led to the definition of the areas of the municipality at risk of flooding which are shown in Fig. 3.6. The depicted areas are the area on the right (2) with high and medium risk of the potential overflow of the Derramador ravine (LLiriet) and the area on the left (1) with medium and low risk of flooding of the beach zone. Following the Master Plan for the Defense against floods in the Marina Baja regions drafted by the Hydrographic Confederation in 2013 (Plan Director de Defensa contra las avenidas en las comarcas de la Marina Baja[5]), the areas of greatest concern in case of flood are concentrated within the urban fabric of Benidorm considering the geographical configuration which is characterized by mountains surrounding the area and the proximity to the sea and where the raverines Barranco Derramador, Barceló, Murtal and Xixo are present.Earthquakes: according to the "Special Plan against Seismic Risk in the Valencian Community", the municipality of Benidorm has a seismic intensity of 7.5 for a return period of 500 years.

Benidorm has a hot semi-arid climate with mild winters and hot summers, it is located between the temperate and tropical zone and it is characterized by significant seasonal variations in rainfall levels and dry summer periods which led to the occurrence of frequent droughts (Linés Escardó 1970; Allen 2001; Trigo et al.

[4] https://politicaterritorial.gva.es/es/web/planificacion-territorial-E-infraestructura-verde/patricova-plan-de-accion-territorial-de-caracter-sectorial-sobre-prevencion-del-riesgo-de-inundacion-en-la-comunitat-valenciana.

[5] https://www.chj.es/es-es/ciudadano/consultapublica/Paginas/PlandirectordefensaavenidasMarinaBajaAlicante.aspx.

2006). Even if Benidorm is one of the driest areas in Spain and one of the most arid in Europe (Romero et al. 1998; Olcina 2004), being surrounded by a relatively rainy proximate mountain environment, its hinterland relies on relatively abundant water resources, essential for the tourist development of the coastline (Martínez-Ibarra 2015); however, the demand posed by the huge number of tourists led to occasional imbalances and periods of crisis; one of the most significant water supply crisis, occurred in 1978. According to Martínez-Ibarra (2015), droughts occurred in Benidorm can be classified as socio-economic droughts or water supply droughts, since they are connected with the touristic demand and the available water supply. Precipitation levels affect the water supply. The study identified different period of water supply drought (1969–1970, 1976–1985 and 1995–2001). In all events, significant precipitations allowed to restore potable water for the population and services. Benidorm strongly depends on rainfall, crucial for ensuring water reserves.

The tourist demand, not only increases water consumption, but as mentioned, it led to an increase of population and housing and to the occupation of risk areas, causing as a consequence the intensification of flood risk. The occupation of the lands, and especially the changes in land use on the Mediterranean coastline occurred in past years, altered both the drainage capacity of the soil and the natural rainwater collectors, many of which have been urbanized, leading to an increase in frequency and severity of flooding (Hernández and Morote 2019).

According to the results of a comparative analysis of different cities located on the Alicante's Coast conducted by Martínez-Ibarra (2017), "the urbanization of small basins and the urban occupation of alluvial planes as well as the loss of traditional functionality of the territory, have amplified the effects of ordinary and extraordinary floods in the analysed areas". The city of Benidorm is characterized by several areas at risk of flooding. Different are the ravines crossing the city, some directed to Poniente Beach and others that flow into Levante Beach. Given the geomorphological conditions of the site, some areas are also at risk of landslides, and due to the lithological characteristics, the water flow have the potential to drag solid material into the collectors, impacting the severity of the potential floods. The ravines that channel into the western and eastern area are characterized by slopes with materials which get easily eroded (Pastor 2004).

According to Pérez-Morales et al. (2018) most of the flood-prone areas that have been occupied are those characterized by a return period of 10 years. The Mediterranean coastal area, is characterized by dense ramification of ephemeral channels. Ephemeral channels are streams which flow only in direct response to a precipitation. They receive very few, sometimes, even no water from springs and they do not get supplied by melting snow or other sources. Ephemeral channels hold water only during and immediately after rain events. These streams are most common in arid and semiarid regions where precipitation is scarce. In order to clarify, the flowing water, i.e. the stream is defined ephemeral, while, channel is the geomorphic bed that conveys the ephemeral flow. Channels may change over time in shape and longitudinal profile due to various processes, but they are not ephemeral. The dry channel of an ephemeral stream, which carries flow for only a limited period of time each year, can dramatically become, due to high-intensity storm, a raging torrent of muddy water and debris in a very short time interval and as soon as the flow diminishes and recedes the channel goes back to its dry state. These typology of natural water channels which are dry for long time, occasionally experience flash floods once extreme precipitation events occur (Hadley 1968). Inopportunely, such features determine a wrong perception of the hazardousness of both the stream and the contingent area (Llasat et al. 2008) encouraging the location of buildings on banks near flood beds influencing the original water channel course.

Pérez-Morales et al. (2018) highlight that flood risk mitigation (structural and non-structural) measures have been implemented in the Alicante and Murcia coastal regions where an increase of flood disasters has been experienced. Thus, the number of fatalities has been visibly reduced. However economic losses have continued to rise, even as a consequence of minor precipitation events.

3.3.1 Recent Plans Delivered by the Municipality

Very recently the municipality of Benidorm has delivered the Municipal Action Plan Against The Risk Of Floods (PAM 2022) and the Climate Change Adaptation Plan (CCAP 2022).

The *Municipal Action Plan for the Risk of Floods* has as primary objective the protection of people, properties and environment exposed to the impact of floods through an operational framework aimed at managing emergency situations making use of public and private resources. The plan identifies critical areas and elements in order to prescribe the most appropriate interventions in case of flood emergencies and it indicates the information and alert procedures for the population. It defines 21 critical areas where the passage of water can be obstructed or sections of communication routes that might be affected by water. Finally, it proposes a sequence of actions to undertake in case of emergency.

The CCAP (2022) represents the commitment of the Municipality of Benidorm to enhance sustainability and resilience in the face of the impacts of climate change already manifested and that, as indicated by climate models, may increase in the coming decades. It is a strategic document developed at local scale compliant with the United Nations Sustainable Development Goals, both the Climate Change Law of the Government of Spain and the Valencian Community, as well as the Valencian Strategy for Climate Change and Energy 2030. The actions and proposals for adaptation detailed for different economic sectors, are proposed on the basis of the evaluation of current and future state of the climatic evolution of the municipality and its potential impact on the water network. The *Plan* focuses on the potential increase of risk due to the increase in extreme precipitation episodes as a result of the local atmospheric effects of Climate Change. The main risk in Benidorm is urban flooding, linked to the existence of ravines crossing the municipal urbanized area. In addition, also forest fires have become a significant concern, due to the changes occurred in land use, in areas where previously there were crop fields, and now transformed into scattered touristic accommodations generating areas of urban-forest interfaces in the vicinity of the Serra Gelada and Sierra Cortina Natural Parks, where currently the forest fire risk is very high. In addition, the increase in maritime storms and the possible rise in sea level can cause major problems in the coastal strip infrastructures.

The methodological basis of the plan to assess the risk generated by climate change is the one established by the Valencian Provincial Council, developed by the Covenant of Mayors for the Climate and Energy of the Valencian Province. The methodology is based on the results of the 5th report of the IPCC (Intergovernmental Panel on Climate Change). The risk level is assessed through a two studies of climate trends in Benidorm (i) Benidorm Climate Projection Study (2021–2100) and (ii) Climate study of Benidorm (past and present). The Plan makes use of the models developed within the World Climate Research Program, and within the framework of the IPCC, which constitute the main international effort to generate global models to determine future climate change scenarios. The Coupled Model Intercomparison Project (CMIP) used is the one related to phase 5 (CMIP5), which serve as the basis for the IPCC report. Considering that the analysis for Benidorm is a local scale analysis and requires local climatic characteristics of the municipality such as series of temperatures and daily rainfall and considering that data derived from Global Climate Models describe global patterns, a downscaling has been carried out to characterize local climate in Benidorm. For this purpose the observed series of temperatures and daily rainfall in Benidorm from 1981 to 2010 have been considered for future scenarios projection. For the missing data in the CCAP, a spatial analysis for their estimation has been carried out by using information from all neighbouring stations. While, two scenarios have been used for the projections: (i) RCP4.5 scenario (projecting a moderate increase in greenhouse gases) and (ii) RCP8.5 scenario (projecting a high greenhouse gases increase, as if mitigation measures did not work leading to an uncontrolled emission). These scenarios have been used to synthetically generate representative series for three periods in the future (2021–2040, 2051–2070 and 2081–2100).

The main climatic hazards to be highlighted as a result of the climate models are:

- Increase in average temperatures, both minimum and maximum, therefore increasing number of tropical nights which can generate a considerable increase in water stress for the vegetation;
- Precipitations do not show a clear trend, they present a slight increase between 2021–2040. The number of precipitations will be reduced, they will be concentrated in fewer episodes with less abundant rain in spring and more abundant in autumn, but with hourly intensification of the rains (torrentiality) causing an increase in runoff;
- Rise in the number and duration of droughts, leading to an increase in water stress of the vegetation, increasing the risk of (i) forest fires, (ii) erosion, and (iii) desertification;
- Less marked seasonality;
- Increase in weather extremes, episodes of torrential rains, severe droughts, heat waves, cold waves, and maritime storms.

As a result of the risk assessment of the plan the main climatic factors that will generate higher risks for the economy, the community and the environment within the municipality of Benidorm are extreme heat, droughts, fires and floods. Different sectors are analysed in the document, some of the results of analysis, selected for their relevance within this work, are here summarized:

- Waste management—the increase in torrential rains and associated runoff can generate an increase in waste to be managed and issues in treatment plants;
- Transport sector—the increase in extreme weather events (torrential rains, strong winds, storms, etc.) will affect the sector conditioning transit routes and will lead to an increased risk of slope instability; climatic conditions, increase in maritime storms and rise in sea level will favour the deterioration of the infrastructures generating higher cost of rehabilitation and maintenance;
- Urban area—increase in energy consumption, exposure of people and assets, torrential rains, strong winds, storms will lead to further deterioration of urban infrastructures and the collection of water resources will be diminished. It will be necessary to include disaster risk management in land use planning avoiding the occupation of risk prone areas and increasing the urban drainage;
- Tourism sector—consequences will be possible changes in the tourist calendar and increase of people in the beach season. The expected increase in extreme meteorological phenomena, maritime storms and the rise in sea level is likely to have a direct impact on tourists and infrastructures (promenades, beaches, marinas). Higher risk for people and assets already located in hazard prone areas has to be expected, requiring to limit future urban developments in the most hazardous areas. Loss and regression of the beaches, due to anthropic alterations such as the modification of ravines channels, or the construction of breakwaters that change the currents and alter the natural water course has also to be expected. A likely increase in storms may further exacerbate the loss of beaches, increasing conservation and maintenance costs.

The Climate Change Action Plan calls for a change with respect to past development, suggesting that "urban development must be carefully considered, avoiding the mistakes of the past that allowed the occupation of risk areas". The plan indicates essential steps:

- Necessity to build protection infrastructures to mitigate the risk of forest fire and hydraulic infrastructures to alleviate flooding and runoff;
- Enhancing warning and alert systems;
- Necessity to address new urban developments to exposure and vulnerability

Following the climate projections for the municipality, 13 lines of actions with different shot- and long-term objectives are proposed. The most relevant to be mentioned, considering the aim of the present work, are:

- Address urban regulations to ensure permeable zones to evacuate water;
- Adaptation of—new and existing—infrastructures avoiding occupation of hazard prone areas, by the (i) implementation of a local plan for flood prevention, (ii)

risk assessment of the critical infrastructures, (iii) development of risk maps, (iv) considering climate change projections in further plans or programmes;
- Promote rehabilitation and urban renewal of the most vulnerable sectors enhancing designs which take into account the effects of Climate Change
- Incorporate criteria for adaptation to Climate Change in the next update of the General Urban Planning Plan of Benidorm (PGOU) which is dated to 1990 and needs to be reviewed considering the impacts of Climate Change and the different hazards threatening the municipality. The modification of the PGOU needs to take into account criteria for adaptation to natural risks and mitigation measures for already urbanized areas at risk.

Those recent plans delivered by the municipality reflect the commitment to enhance and improve the current situation in Benidorm with a view to increase its resilience in the face of the impacts of climate change. However, two main areas of improvement can be still envisaged. No plan specifies how the tourism sector should be differently managed in the future. Those plans insofar lack a focus on tourists as vulnerable as a consequence of language barriers and lack of familiarity with the environment with its hazards and risks. In addition, the 13 lines of actions proposed in the Climate Change Action Plan are comprehensive and ambitious, but implementation time and the amount of money allocated seem not reasonable. The changes and improvements that they are seeking cannot be implemented in such a short time, they require proper analysis, understanding, sharing and involvement of different groups of the community in order to deliver proper efficient and effective mitigation measures and strategies that can support the enhancement of resilience.

3.4 Data Collection and Processing

Both qualitative and quantitative data have been used for the development and application of the methodology. This allowed to exploit the resourcefulness of different data sources in order to construct a robust information and knowledge research ground.

Several data sources have been explored and investigated to retrieve as much as possible relevant records or information. These data sources include amongst others direct observations, official documents and administrative delivered by the municipality, archival records, maps and cartographies, newspaper articles, reports, research contributions, academic papers and so on; most of them available only in Spanish language. The experience abroad at the University of Alicante, the meetings and discussions held in Benidorm, as well as those held online have been a further source of information. The interaction with researchers, local stakeholders and site visits have allowed to construct an inventory of data and information related to the case study which have been then elaborated and used to implement and test the proposed methodology.

Having drawn both data and information from heterogeneous sources, data have been processed and managed in order make them usable for the research.

The set of data of which this research makes use can be divided in two type of inputs which have been handled and processed with different approaches: (i) data and information collected during the exploration of different sources, such as through the interaction with local stakeholders; and, (ii) data provided by the Spanish Insurance Compensation (Consorcio de Compensacion de Seguros, CCS) of the compensation amount reimbursed due to flash floods damages occurred between 1996 and 2021 in the Alicante Province.

Data and information collected through different sources, i.e. of data of different formats (visual data, textual data, spatial and georeferenced data) have been handled and processed in different ways to develop text sections to reconstruct narratives, but also to build tables and elaborate maps with punctual information given the refined level of detail found for some specific areas. Precisely those data allowed to construct properly the case study, i.e. the urban development characteristics, the consequent dynamics of exposure and vulnerability increase, such as the flood events history of Benidorm. Also those sources allowed to design the three stress testing presented in Chap. 4, metrics and their scores have been allocated in function of the information collected through those different sources. For instance the analysis of maps and cartographies as well as administrative documents allowed to implement the first stress testing; the subsequent two stress testing have been based on data collected through direct observations, official documents, newspaper articles, research contributions about affected people, damages to infrastructures and details about their loss of services, as well as insights about the emergency management, the environmental damages, especially occurred on the beach areas, in addition to more detailed (qualitative and quantitative) data about the extent of damages on specific assets (school, touristic facilities and road elements).

The numerical dataset provided by the CCS constitutes a valuable set of observations through which it has been possible to reconstruct the damage dynamics of the case study. Such data are referred to compensations amount reimbursed in case of damages due to flash floods occurred between 1996 and 2021 in the Alicante province. They have been provided at the municipal level for the whole province and at the lowest granularity level for Benidorm, which is the Postal Code. Having the attribute of the postal code permits to give a spatial meaning to the records and allows to detect the most damaged areas within the city, by years or alongside the whole time range for instance.

3.5 Flood Damage and Loss Data Analysis

The CCS founded in 1954, is the Spanish National Insurance System coverage for 'extraordinary risks', which includes natural phenomena and events of political and social nature, that may affect people and properties. It is a public corporate entity related to the Ministry of Economic Affairs and Digital Transformation of Spain and provides compensations due to occurred damages on insured assets and people holding an insurance scheme. The CCS has specific legal characteristics and is subject to laws governing private sector firms (Barredo et al. 2012), it collects data on direct physical damage (including people) and loss of profit with the purpose of compensation in terms of new replacement value. The consortium does not collect data related to non-insured assets. Available data in CCS databases are mostly related to the compensations amount reimbursed, data are referred to damage type (personal, property, loss of profit) and type of property (housing, shops and stores, business offices, industry, public infrastructure, vehicles). The price of the cover offered by CCS is a surcharge on the sum insured in policies issued by private insurance firms, which collect premiums and then the surcharge is transferred to CCS. Entitlement to compensation for one of the extraordinary events covered by the scheme arises only where the injured party has already taken out an insurance policy, taking an insurance policy with an insurer automatically carries the coverage against extraordinary risks. Insurance companies collect a surcharge along with the premium and transfer it to CCS. CCS surcharge is the result of applying a specified rate to the sum insured under the policy. The rate differs according to the type of value covered but is applied generally for the whole of Spain and for all risks included under the scheme without distinction, and is independent of the degree of exposure. Entitlement to indemnification does not take into account the number of claims nor the geographical area, or the total damage arising in the claim but is only related to the occurrence of damages due to the contemplated extraordinary risks. The Real Decreto 300/2004 is the Extraordinary Risk Insurance Regulation currently in force. In order to better understand main features of the data-set handled in this work, it is essential to highlight some main points: (i) an extraordinary risks cover is compulsory in the policies covering damage from natural events such as floods; (ii) before 1986 an official declaration of catastrophe zone was required for a flood to be defined as extraordinary and then be covered by CCS, from that year the policy changed the definition of "extraordinary risk" so that it was not compulsory anymore an official declaration and flood impacts have been automatically covered by CCS; (iii) from 2004 losses related to business interruption as result of direct damage have been included. In addition, when summing the losses over the years, even when adjusted, it is necessary to be aware of the fact that the variation of the losses claimed and the surcharges vary over time due to the changes of both the value and the number of the assets insured due to the insurance penetration (Barredo et al. 2012). Considering an analysis of flood loss data in Spain at the municipal level carried out by the CCS (2021), indemnities for flood losses, including coastal floods, reach up to 69% of all compensation covered for the "extraordinary risks" insurance scheme in the past 25 years, since 1996.

Data provided by the CCS, in excel format, contain: date, type of event, type of damaged asset and the relative compensation. They have been filtered for the area of interest; specifically considering that they are aggregated at the level of the postal code, the following codes of Benidorm city have been used to select the records: 03500, 03501, 03502 and 03503. Those areas are spatially represented in Fig. 3.7.

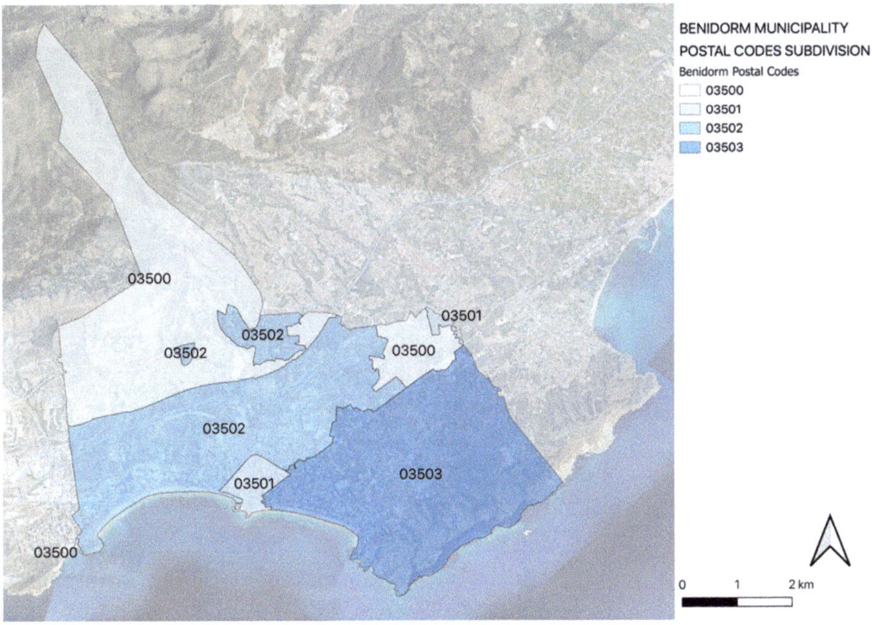

Fig. 3.7 Representation of Benidorm postal codes areas

The total records obtained by the filtering are 960. Those records refer to the time span ranging from 1996 to 2021.

The compensation records, considered as economic losses, are referred to the year of occurrence of the event. Therefore, the initial value (i.e. nominal value) referred to the year of occurrence of the event is adjusted through a coefficient, the Consumer Price Index, in order to have comparable values across time. The Consumer Price Index (CPI) is a statistical measure of the evolution of the prices of goods and services consumed by the population residing in family homes in Spain. The set of goods and services that make up the shopping basket is basically obtained from household consumption and the importance of each of them in calculating the CPI is determined by said consumption. In this case the year of reference that has been used is the year 2021.

Table 3.1 contains all the years for which the CCS have claimed pay-outs between the years 1996 to 2021 in the city of Benidorm with the sum of the total compensation per year, the two last columns report respectively the nominal values and the adjusted one having 2021 as a reference year.

From the initial analysis, it is evident that the year that has recorded the highest damages is 2007, followed by 1997 and 2009.

The following Fig. 3.8 represent the spatial distribution of damages that have occurred in the time range between 1996 to 2021 per postal code of the city.

The following Table 3.2 shows the five years for which the highest damages have been recorded in Benidorm as a function of the total claims compensated by the CCS.

Table 3.1 Sum of the CCS compensation per year for Benidorm

Year	Nominal value	Adjusted value (2021)
1996	€ 932	€ 1.583
1997	€ 388.724	€ 647.473
1998	€ 391	€ 642
2000	€ 64.610	€ 99.173
2001	€ 7.640	€ 11.418
2002	€ 162.768	€ 233.915
2003	€ 9.415	€ 13.187
2005	€ 5.399	€ 7.067
2006	€ 212.798	€ 271.195
2007	€ 3.878.627	€ 4.743.772
2008	€ 33.278	€ 40.139
2009	€ 347.346	€ 415.633
2010	€ 39.536	€ 45.930
2011	€ 116.153	€ 131.778
2012	€ 70.555	€ 77.789
2013	€ 1.079	€ 1.186
2014	€ 156.393	€ 173.650
2015	€ 135.020	€ 149.919
2016	€ 34.830	€ 38.064
2017	€ 131.066	€ 141.679
2018	€ 40.714	€ 43.489
2019	€ 58.753	€ 62.260
2020	€ 8.949	€ 9.531
2021	€ 4.989	€ 4.989

The spatial analysis of recorded data shows that the area that has received more pay-outs is the 03500 postal code area with almost 4 MLN EUR compensated damages, followed by 03503 area with more than 2 MLN EUR total compensation. Table 3.3 reports the total compensation per each postal code in adjusted values with reference to 2021.

Focusing the analysis on the most damaging event, i.e. the flood occurred in October 2007, a more refined analysis has been carried out to evaluate the sectorial category that has suffered the highest level of damages. Table 3.4 reports the value of compensation per category. Commercial activities suffered the highest levels of damage as expected given the touristic vocation of the area of study, followed by residential sector, offices, vehicles and industry.

The following Fig. 3.9 allows to understand where most of the damage has been recorded in the city according to the postal code subdivision.

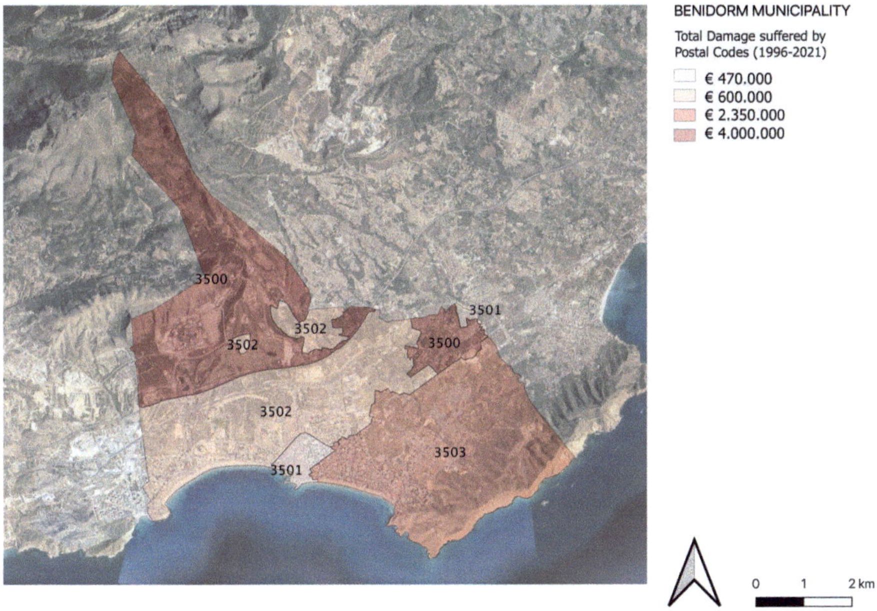

Fig. 3.8 Spatial distribution of the total damage (1996–2021)

Table 3.2 List of the five years in which the highest values of compensation has been recorded in Benidorm

Year	Nominal value	Adjusted value (2021)
2007	€ 3.878.627	€ 4.743.772
1997	€ 388.724	€ 647.473
2009	€ 347.346	€ 415.633
2006	€ 212.798	€ 271.195
2002	€ 162.768	€ 233.915

Table 3.3 Total compensation in the time range 1996–2021

Postal codes	Total compensation (1996–2021) adjusted values
3500	€ 3.960.791
3501	€ 470.490
3502	€ 600.862
3503	€ 2.345.549

The compensations amount reimbursed due to occurred damages are referred only to the insured assets, the consortium does not collect data related to non-insured assets. The CCS coverage for "extraordinary risks" is automatically contracted when underwriting an insurance scheme that contemplates property loss, personal injury

Table 3.4 Most damaged categories in 2007

Category	Tot. losses (nominal value)	Tot. losses (adjusted values-2021)
Commercial activities	€ 1.551.700	€ 1.898.000
Residential	€ 1.293.300	€ 1.581.800
Offices	€ 480.600	€ 588.000
Vehicles	€ 277.000	€ 339.000
Industry	€ 271.000	€ 331.000

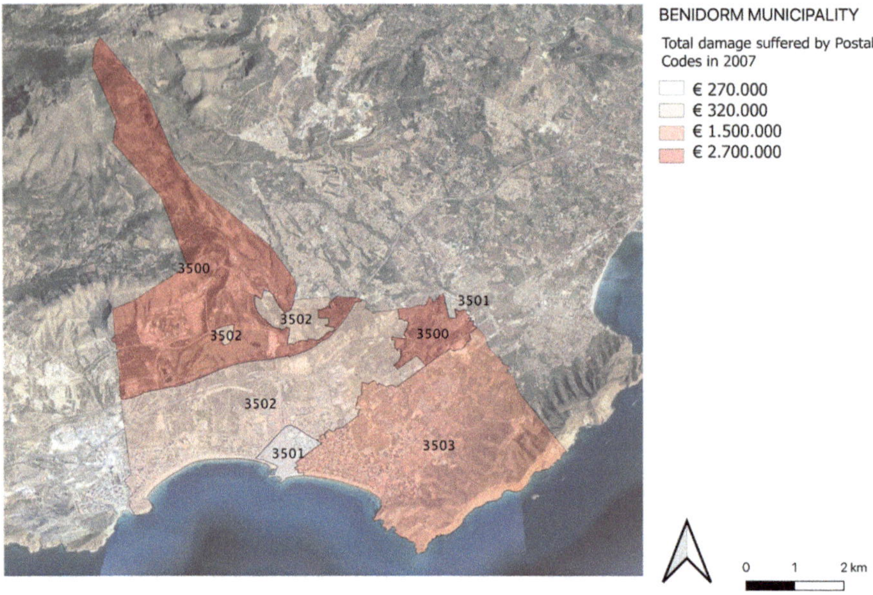

Fig. 3.9 Distribution of the damage occurred in 2007

and business interruption with any insurance operating in Spain. Therefore, knowing the level of insurance penetration in the years and knowing the percentage uninsured assets, especially homes would allow to perform a more comprehensive analysis. Following the discussion with the representative of the CCS, this portion of data analysed could be under-representative considering that the case study analysed is a tourism destination in which the second homes rate is high and the percentage of insured homes might present particular specificities. In addition, as already described in the introductory section of the damage and loss data analysis, when summing the losses over the years, even when adjusted, it is necessary to be aware that from 2004 losses related to business interruption as result of direct damage have been included and of the fact that the variation of the losses claimed and the surcharges vary over time due to the changes of both the value and the number of the assets insured (Barredo et al. 2012).

References

H.D. Allen, in *Mediterranean Ecogeography Ecogeography Series* (Harlow, 2001)

K.O. Arnstberg, I. Bergstrom, No place like second home: weekends, holidays, retirement and urban sprawl, in *Urban Sprawl in Europe: Landscapes, Land-Use Change and Policy* (2007), pp. 163–80

I. Baidal, J.A.I. Sánchez, I.R. Rebollo, The evolution of mass tourism destinations: new approaches beyond deterministic models in Benidorm (Spain). Tour. Manag. **34**, 184–195 (2013)

J.I. Barredo, D. Saurí, M.C. Llasat, Assessing trends in insured losses from floods in Spain 1971–2008. Nat. Hazard. **12**(5), 1723–1729 (2012)

C. Bellet Sanfeliu, The evolution of urban planning in medium-sized Catalan cities (1979–2019). Urban Sci. **5**(2), 36 (2021)

CCAP, in *Climate Change Adaptation Plan* (2022). Available at: https://contenidos.benidorm.org/sites/default/files/descargas/2022-05/PLAN%20DE%20ADAPTACIÓN%20ANTE%20EL%20CAMBIO%20CLIMÁTICO%20DE%20BENIDORM_OCT21_IC.pdf

CCS, in *Analysis of flood loss data in Spain at the municipal level. Number 14, Spring 2021* (2021). Available at: https://www.consorsegurosdigital.com/en/numero-14/front-page/analysis-of-flood-loss-data-in-spain-at-the-municipal-level

P. M. Ciriquián, J.O. Meyer, Algunas intervenciones en la costa mediterránea de Alicante: tres propuestas, tres ideologías, tres actitudes, in *Arquitectura moderna y turismo: 1925–1965: Actas IV Congreso Fundación DOCOMOMO Ibérico* (2004), pp. 225–230

P.M. Ciriquián, A. Nolasco-Cirugeda, L. Serrano-Estrada, Estrategias de ocupación territorial en áreas turísticas consolidadas de la Costa Blanca (España). Boletín De La Asociación De Geógrafos Españoles **76**, 430–454 (2018)

L. Corbusier, in *Quand les cathédrales étaient blanches* (Plon, Paris, 1937). Available in English at: https://monoskop.org/images/f/fc/Corbusier_Le_When_the_Cathedrals_Were_White.pdf

M.M. Del Castillo. *Benidorm. Esplanades And Ordinary Urbanities* (2014). Available at: https://www.academia.edu/10465959/Benidorm_Esplanades_And_Ordinary_Urbanities

L. Domínguez Martínez, P. Martí Ciriquián, A. Nolasco-Cirugeda, *Turismo Residencial de Extranjeros en la Costa Blanca: su Manifestación Territorial y Estudio de Casos Representativos* (2016)

A.T. Edwards, A further criticism of the Garden City Movement. The Town Plann. Rev. **4**(4), 312–318 (1914)

J. Fariña, J.M. Naredo, *White paper on sustainability in Spanish urban planning* (Ministry of Housing, Madrid, 2010)

L. Fernández-Galiano, in *Benidorm en positivo* (El País, 2002)

R.M. Fraguell, C. Martí, J. Pintó, G. Coenders, After over 25 years of accrediting beaches, has Blue Flag contributed to sustainable management? J. Sustain. Tour. **24**(6), 882–903 (2016)

R.F. Hadley, Ephemeral streams, in *Geomorphology. Encyclopedia of Earth Science* (Springer, Berlin, Heidelberg, 1968). https://doi.org/10.1007/3-540-31060-6_108

P. Hall, *Cities of Tomorrow: An Intellectual History of Urban Planning and Design since 1880* (John Wiley & Sons, 2014)

M. Hernández, A.F. Morote, The use of rainwater in Alicante (southeast Spain). A new urban approach to urban water management. UPLanD-J. Urban Plann. Landscape Environ. Design **4**(1), 53–66 (2019)

H. Lefebvre, Toward an architecture of enjoyment, in L. Stanek (ed.), : (University of Minnesota Press, London, England, 2014)

A. Linés Escardó, The climate of the Iberian Peninsula. Clim. Northern West. Europe **5**, 195–226 (1970)

M.D.C. Llasat, L. López, M. Barnolas, M. Llasat-Botija, Flash-floods in Catalonia: the social perception in a context of changing vulnerability. Adv. Geosci. **17**, 63–70 (2008)

W. Maas, J. van Rijs, N. de Vries, *Costa Ibérica. Hacia la ciudad del ocio* (Actar, Barcelona, 2000)

E. Martínez-Ibarra, Climate, water and tourism: causes and effects of droughts associated with urban development and tourism in Benidorm (Spain). Int. J. Biometeorol. **59**(5), 487–501 (2015)
E. Martínez-Ibarra, Geographical analysis of flash floods in tourism areas in the north of the province of Alicante. *Papeles de Geografia* **63**(Special Issue), 65–81 (2017)
T. Mazón, E. Delgado, J.A. Hurtado, El éxito de un destino turístico: el Benidorm de Mario Gaviria. Revista da Casa da Geografia de Sobral **14**(1), 81–95 (2012)
MTP. *Plan Territorial frente a Emergencias Municipio de Benidorm* (2019). Available at: https://contenidos.benidorm.org/sites/default/files/documentos/ptme_benidorm_2019lpd.pdf
A. Nolasco-Cirugeda, P. Martí, G. Ponce, Keeping mass tourism destinations sustainable via urban design: the case of Benidorm. Sustain. Dev. **28**(5), 1289–1303 (2020)
J. Olcina Cantos, *Spatial planning processes, territorial planning law and flood risk in the region of Valencia (Spain)* (Risks Challenging Publics, Scientists and Governments, 2010), p.191
J. Olcina Cantos, M. Hernández Hernández, climate and water on the Spanish mediterranean coast: challenges for the future. Geograph. Mediterranean Europe 121–143 (2021)
A.G. Olcina, La región climática del sureste ibérico, in *Aridez, salinización y agricultura en el sureste ibérico* (Editorial Universitaria Ramón Areces, 2004), pp 13–35
J.S. Palencia Jiménez, E. Gielen, Y. Pérez Alonso Resilient urban morphology to the flood risk in the final stretch of the Girona River (Alicante province), in *24th ISUF International Conference. Book of Papers* (Editorial Universitat Politècnica de València, 2018), pp. 995–1003
PAM. Municipal Action Plan Against The Risk of Floods In Benidorm (2022). Available at: https://contenidos.benidorm.org/sites/default/files/descargas/2022-03/1_pdfsam_pam_inundaciones_benidorm_14022022_lpd.pdf
R.Z. Pastor, Delimitación de los sectores con riesgo de inundación en los municipios costeros de la Marina Baja: Análisis de sus causas y posibles actuaciones de carácter preventivo. in *Aguaceros, aguaduchos e inundaciones en áreas urbanas alicantinas* (Servicio de Publicaciones, 2004), pp. 253–288
A. Pérez-Morales, S. Gil-Guirado, J. Olcina-Cantos, Housing bubbles and the increase of flood exposure. Failures in flood risk management on the Spanish south-eastern coast (1975–2013). J. Flood Risk Manag. **11**, S302–S313 (2018)
J.M. Pérez, Urban planning system in contemporary Spain. Eur. Plan. Stud. **15**(1), 29–50 (2007)
J.I. Prieto-Arranz, M.E. Casey, The British working class on holiday: a critical reading of ITV's Benidorm. J. Tour. Cult. Chang. **12**(1), 68–83 (2014)
G.A.M.S. Rodríguez, Urban growth in Benidorm according to the documents for works (1950–1970). Investigaciones Geográficas **30**, 119 (2003)
R. Romero, J.A. Guijarro, C. Ramis, S. Alonso, A 30-year (1964–1993) daily rainfall data base for the Spanish Mediterranean regions: first exploratory study. Int. J. Climatol. J. Royal Meteorol. Soc. **18**(5), 541–560 (1998)
E. Sánchez-Almodóvar, J. Olcina-Cantos, J. Martí-Talavera, A. Prieto-Cerdán, A. Padilla-Blanco, Floods and adaptation to climate change in tourist areas: management experiences on the coast of the Province of Alicante (Spain). Water **15**(4), 807 (2023)
M.A Soto, *Cambios en el paisaje 25 años después del incendio forestal de Terra Mítica.* Available at: https://archivo-es.greenpeace.org/espana/es/Blog/incendio-terra-mitica/blog/60008/. Last access 21 September, 2022 (2017)
Plan especial riesgo incendios, Generalitat Valenciana. Available at: https://prevencionincendiosgva.es/Documents/PlanesVigilancia/legislacion/302%20Decreto_163-98_plan_especial_riesgo_incendios.pdf. Last access 22 November, 2022 (2010)
R. Trigo, E. Xoplaki, E. Zorita, J. Luterbacher, S.O. Krichak, P. Alpert, ... A. Mariotti, Relations between variability in the Mediterranean region and mid-latitude variability, in *Developments in Earth and Environmental Sciences* (vol. 4, Elsevier, 2006), pp. 179–226
J.F. Vera, in *Turismo y urbanización en el litoral alicantino [Tourism and urbanization in the coast of Alicante]* (Instituto de Estudios Juan Gil-Albert, Alicante, Spain, 1987)
P. Zaragoza, *Así será Benidorm* (Tipografía artística, Madrid, 1955)

Chapter 4
Stress Testing the Resilience of the City of Benidorm and Its Planning System

Abstract This chapter shows the application of the methodology described in Chap. 2 to the city of Benidorm. Damage and loss data allowed to build the stress tests applied to the city and its urban planning, which results are framed within the resilience matrix and discussed, assessing to what extent the spatial development of the city and its planning system favor or diminish its overall resilience.

4.1 Using Post-disaster Damage and Loss Data

The selection and analysis of damage ad loss data is a crucial step of the methodology development and application. The search of damage ad loss data and the analysis process on the available data allowed to implement the methodology proposed which combines two different tools: stress testing and resilience matrix. Damage and loss data allowed to build the stress tests and to make considerations about resilience; they allowed to understand the most stressed components of the system and to reconstruct its vulnerabilities.

This empirical work is intended to be valid for analysing damage and loss data produced and shared in different formats, related to the impact of different hazards to multiple societal sectors. The type of analysis to carry out varies according to the type of data available, the damage and loss data analysis is not here presented as a method defined univocally which can be replicated identically, but rather as a conductor line, since the input data can vary from case to case.

Specifically, this work proposes an analysis which makes the best use of the available data.

The first set of data refers to *post-flood damage data*, where "post" indicates the time of the data collection. Damage data refers to the impacts of floods on different exposed sectors.

The analysis is carried out with the aim to understand: "What do the available damage and loss data say about the territory under consideration? What is possible to learn about the system through such evidence?". This analysis varies according to the type of data available, in the ideal case damage and loss data should be collected after

A. Faiella, *Learning from Disasters to Enhance Resilience of Urban Complex Systems*, PoliMI SpringerBriefs, https://doi.org/10.1007/978-3-032-07356-3_4

all events occurred due to different hazards (floods, wildfires, landslides, earthquake etc.) and for all societal sectors at the asset level including the temporal stamp (IDEA; LODE; De Groeve et al. 2013; Faiella 2020; Faiella et al. 2022) so that proper analyses, for instance a forensic analysis, could be carried out to acquire lessons learned after a disaster with the aim to support spatial planning practices in risk prone areas.

The proposed methodology, which contemplates the stress testing and its implementation in the resilience matrix, not only allows to embrace the complexity of contemporary systems, but as a standardized but flexible procedure it allows also to overcome issues related to the heterogeneity of damage and loss data analysis outputs; an expected condition, intrinsically related to the variability of the input data of such analysis (textual descriptions, maps, images, economic value of reimbursements, videos, tabular data and so on).

4.2 Stress Testing and Resilience Matrix Applied to Benidorm

In this chapter the methodology described in Chap. 2 is applied to the city of Benidorm. The results of the stress testing the city and its urban planning are discussed, assessing to what extent the spatial development of the city and its planning system favor or diminish its overall resilience.

Three stress testing scenarios have been applied on the urban system of Benidorm.

Specifically the three applied stress testing take into account three different temporal stressors referred to present, past and future; respectively as follow:

- Stress Testing A: Current data availability stress testing. Evaluation of the data quality and of their accuracy to satisfy the Sendai Indicators calculation.
- Stress Testing B: Evaluation of urban quality and resilience of the complex urban system of Benidorm when applied the 2007 event.
- Stress Testing C: Applying the Climate Change impacts and Urban Growth projections on the urban system of Benidorm

Stress Testing A is based on the experience gained during the research path, aiming at delivering an evaluation of the system performance when it is stressed by the "search of accurate data" necessary for the implementation of our methodology and to satisfy the Sendai Indicators' requirements. While Stress Testing B takes into account the most damaging event occurred in terms of higher economic losses recorded, which is the flash flood occurred in 2007, and Stress Testing C takes into account the projections of climate change impact on the city following the analysis of the CCAP (2022). This methodological choice is driven by the necessity to evaluate properly the whole system behaviour when heterogeneous stressors are applied.

In line with the aim of the work of developing an operational and effective methodology, the stress testing procedure and the following resilience matrix construction,

General profile	From agricultural village to mass-tourism destination
Urbanized area	From 21 hectares in 1956 to 867 hectares in 2019
Population	From 6000 inhabitants in 1960 to 70.000, reaching 215,000 inhabitants/day in high season
Hazard	Urban flash floods (ephemeral channels)
Criticalities	Urban growth in flood prone areas and scarcity of water resources especially in high season

Fig. 4.1 Most salient characteristics of the defined complex system

where the performance scores of the system under the stress load are allocated, follow the same framework for the three stress tests that have been applied (see Figs. 2.5 and 2.6). The metrics identified to track and determine along the stress testing process system behaviour and the relative scores, ranging from 1 to 4, meaning—very low, low, moderate and high—performance level, are assigned in function of the understanding of the system gained throughout the whole research path.

For all of the three stress testing, A, B and C, the Step 1: "Definition of the System" is the same providing the analysis of the city of Benidorm as a system in its. Figure 4.1 shows the most salient characteristics of the defined complex system.

4.3 Stress Testing A: Evaluation of the Data Quality and of Their Accuracy to Satisfy the Sendai Indicators Calculation

Stress Testing A Testing is carried out by searching for accurate and good quality data to implement the proposed methodology and to understand their usability and reliability to assess the Sendai Framework Indicators.

4.3.1 Definition of the Stressor

The analysis of the urban system of Benidorm, i.e. search of meaningful data, consists in a stress applied to the case study per se. During the phase of analysis of the area of interest, especially in the "Step 1: definition of the system", the procedure implied the application of a stressor on the system, in terms of "search for available data".

This essential step of "data search", carried out to accomplish our work and to implement the developed methodology, generated an intrinsic stress-testing of the performance of the system when queried. The procedure to access data to properly investigate and define the case-study allowed to have a broader view of the system informational resources and to evaluate its performance in terms of data collection and data governance. In addition, collected data have been evaluated in terms of consistency with the Sendai Indicators' requirements.

Collecting data is an important step towards improved prevention since it allows to identify criticalities and understand the system's behaviour in case of disruptions. The collection of data in this application doesn't provide only a context, or a starting point, for analysis, but also a benchmark for current performance problems. Searching, analysing and reviewing such data served to further define the other two stress tests as well.

4.3.2 Application of the Stressor and Evaluation of the Effects

The stress applied consists in querying the system while searching for data deemed necessary for a comprehensive risk and resilience assessment. In order to perform both risk and resilience assessment it is it crucial to evaluate and benchmark the baseline state of the system (Cutter et al. 2010) and the lack of consistent quality data is an issue that hampers the resilience of a system. Scarcity of available data means scarce knowledge of the potential phenomenon that might occur in the area of interest and of its impacting dynamics.

Stress Testing A consists in querying the system to provide useful data and information to carry out the baseline assessment. In the intention to answer the main question that arises to apply the stressor "Is the data storage/availability sufficient for different types of applications?", the following issues have been addressed:

- Which data can be accessed?
- Are those referred to the local scale of interest?
- Do regulations address the processing of necessary data considering the hazardous characteristics of the area?
- Do previous researches and works carried out explore the significant phenomenon of the area?

This Stress Testing allowed to identify performance problems of the system.

Most of the available data come from the literature review carried out on previous research works and on the very recent plans delivered by the municipality, such as the MTP (2019), the PAM (2022) and the CCAP (2022).

The Territorial Plan against Emergencies highlights the exposure of the municipality to multiple hazards such as: floods, forest fire, landslides, earthquake and droughts; but it doesn't provide risk analysis nor the evaluation of hazard, exposure and/or vulnerability. It is a very descriptive document and does not contemplate specific measures.

The municipality has, and just very recently—mid 2022—delivered the PAM, an operational tool which delineates actions and strategies to carry out in case of floods. The CCAP (2022) highlights clearly the need to develop and carry out proper analysis also for seismic and forest fire hazards and to enhance local plans for the prevention of floods and for the urban drainage to support jointly both enhanced spatial planning and emergency management.

In general, the municipal plans deliver useful maps, however row data are not available. The plans carry out some preliminary analysis, still there are no specific studies regarding vulnerability, neither analysis in terms of quantification or monetary (quality) evaluation of exposure; for instance there are no data about exposed people, number of people living in flood prone areas, layer of population intersected with flood data. The same for residential, industrial buildings or critical facilities. The CCAP (2022) provides some maps, but only of hotels located in flood and forest fire prone areas. Besides, the aforementioned plans do not indicate flooded areas based on local hydrological models, nor information about water velocity or estimated duration of the floods. There are no considerations about seasonality and number of tourists, limitations and increased density.

The Valencian Cartographic Institute (Institut Cartogràfic Valencià[1]) provides some vectorial data, however the availability is limited, and specific data at local scale are not available.

The MTP (2019) does not contemplate specific measures, it is a very descriptive document. While the PAM (2022) addresses some strategies and actions, but since it is a very recent document it is not considered in this stress testing.

4.3.3 Resilience Assessment

The procedure, considered as the stress solicitating the system helps identify areas that might need investments and once fulfilled the overall resilience might systematically improve because data to validate and calibrate models and strategies will be available and the overall situation might improve.

Table 4.1 illustrates the process carried to construct the resilience matrix for Stress Testing A.

The table contains in the first column the identified critical functions of the whole system to be evaluated when the stress i.e. the data quality analysis is applied.

For instance, in the first row the Municipal Emergency Plan is taken into account as a metric to be assessed. Operationally the performance score is assigned in function of the question "are the data available of a good quality to implement the methodology?". Considering the very qualitative description of the potential hazards that

[1] https://icv.gva.es/va/.

Table 4.1 Metrics and score of stress testing A to implement within the resilience matrix

Stress testing A—data availability and quality stress testing								
Parameters considered		Performance score				Final score	Resilience matrix allocation	References
		Very low [1]	Low [2]	Moderate [3]	High [4]			
Municipal emergency plan						2	Prepare/information	MTP (2019)
Municipal action plans and protocolsfor different hazards						1	Prepare/information	PAM (2022)
Multi-hazard management plans						1	Absorb/physical	PAM (2022), CCAP (2022)
General Urban Development Plan (PGOU) dating to 1990						1	Prepare/social	Chapter 3
Flood risk mapping	Hazard					2	Prepare/information	PAM (2022)
	Exposure					1	Prepare/economic	PAM (2022)
	Vulnerability					1	Prepare/social	PAM (2022)
Risk mapping	Hazard (earthquake, forest fire, landslide)					1	Prepare/information	PAM (2022)
	Exposure					1	Prepare/economic	PAM (2022)
	Vulnerability					1	Prepare/social	PAM (2022)
Studies and analysis at local scale						2	Adapt/information	Chapter 3

(continued)

Table 4.1 (continued)

Hydrological models (water velocity, floods duration)						1	Prepare/information	Chapter 3
Available raw data						2	Prepare/social	Valencian Cartographic Institute
Tourists	Knowledge of the local environment (hazards and self-protection measures)					1	Absorb/social	Aznar-Crespo et al. (2020), Matyas et al. (2011)
	Denote risk conditions during promotion and sale for tourism					1	Prepare/social	Aznar-Crespo et al. (2020)
	Inclusion in the existing municipal Plans					1	Recover/economic	MTP (2019), PAM (2022), CCAP (2022)
Damage and loss data						3	Prepare/information	CCS
Compliance with the computation of the Sendai framework indicators						3	Adapt/information	CCS

could impact the city in the Municipal Emergency Plan and that it doesn't provide neither a risk nor a multi-hazard assessment, the result is allocated in the second slot with a score equal to 2, meaning low performance score. Then, the position within the resilience matrix is assigned; in this case in the intersection between the prepare phase and information domain.

The absence of plans and protocols for other types of hazards, rather than floods, is evaluated by assigning a very low score of performance to that second metric, allocated in the same cell of the resilience matrix as for the first metric.

Given that the municipal General Urban Development Plan dates back to 1990 and does not make reference to any potential hazardousness situation and only scanned paper maps are available for consultation, a score equal to 1 (very low performance) has been assigned and allocated in the prepare/social intersection of the resilience matrix. The same procedure is followed for the other metrics and the score is always assigned in function of the results obtained and considerations carried out during the definition of the system under analysis. As described in the methodological chapter (Chap. 2), when more than one performance score is allocated in the same cell of the matrix, as in this case, at the end they are joint in the final matrix with an average (in Fig. 4.2 for the cell prepare/information the final value is the result of the following computation *prepare/information = sum of the performance scores allocated in the same cell/divided by the number of parameter considered* = *(2 + 1 + 2 + 1 + 1)/*5).

A precise consideration of the flood risk has been done, regarding the hazard a low score has been assigned, while for exposure and vulnerability very low, as most flood risk maps do not take into account yet exposure and vulnerability properly. Very low values have been assigned to risk mapping in general, considering the lack experienced in the search for hazard data (earthquake, forest fire, landslides) at the

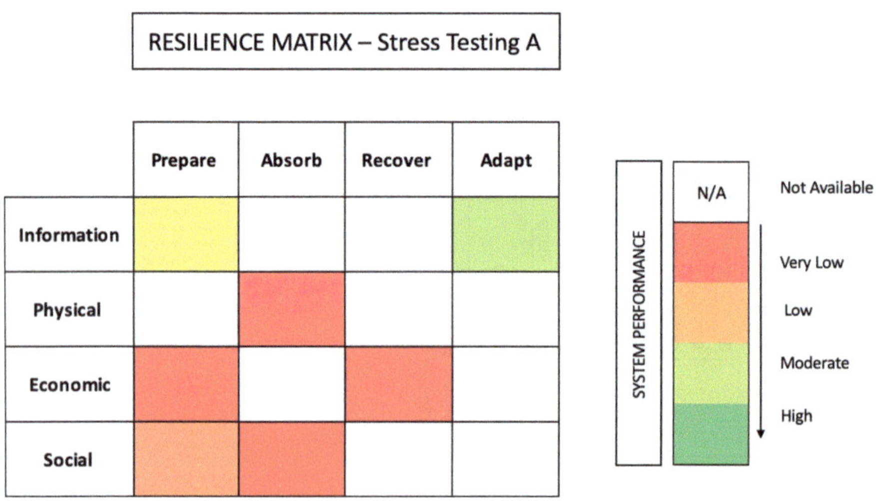

Fig. 4.2 Resulting resilience matrix of stress testing A

municipal or even regional level. Also missed consideration of the tourists, a very significant portion of the population, especially in the high season has been assessed as a negative condition.

In addition, Stress testing A regards also the evaluation of the consistency of the available data and information to satisfy the Sendai Indicators' requirements. The general set of available data analysed does not permit to assess the indicators established to measure progress in achieving the global targets of the Sendai Framework. Only the indicators of Target C, which aims at monitoring the direct economic loss attributed to disasters accounted in relation to global gross domestic product, could be computed by making use of the data provided by CCS. Specifically, C-1 compound indicator related to direct economic loss attributed to disasters in relation to global gross domestic product, C-3 direct economic loss to all other damaged or destroyed productive assets attributed to disasters., C-4 direct economic loss in the housing sector attributed to disasters. Such analysis has been carried out within the user corner of the Risk Data Hub. The performance of the system is therefore assessed with a score equal to 3, meaning moderate performance score. The score is assigned according to the use that can be done to satisfy the Sendai Indicators requirements; the CCS data has been advantageous and allowed to make some detailed analysis for Target C, but the specific performance is not evaluated with a high score because the available dataset is the only available at the system level and does not provide sufficient granularity at the sub-city level (neighbourhoods).

The final results obtained are represented in the following Resilience Matrix (Fig. 4.2), where the colours from red to green represent a gradually a more resilient behaviour, as showed in the legend on the right side.

After applying the stress testing A, a general very low to low performance of the system is recorded as showed in the resilience matrix.

This first stress testing allows to make an assessment of the system in relation to the data availability and of the quality of the existing ones. Such initial type of stressor applied becomes a fundamental step allowing to understand the knowledge background of the entire system.

Between the prepare, adsorb and recover phases the general system performance is characterized by a low score. A slight improvement is evident in the prepare/information and the adapt/information intersection of the matrix. These results are given respectively by the damage and loss data metric which allowed to assess the system performance with a moderate score given the CCS dataset; and, by the moderate system performance score related to the compliance with the computation of the Sendai Framework Indicators metric.

Therefore, this result clearly shows the lack of available fundamental data at the system level to support spatial and urban planning, considered as mitigation strategy to reduce risk, by acting not only on hazard but especially on exposure and vulnerability, which as seen, play a crucial role in the city of Benidorm and to address and enhance resilience.

4.4 Stress Testing B: Evaluation of Urban Quality and Resilience When Applied the 2007 Event

Stress Testing B is based on a scenario constructed taking into account the current urban configuration of Benidorm and applies the 2007 flash flood as stressor, according to the analysis performed on the CCS loss dataset it is the event that caused most economic losses than any other event. This stress test makes direct use of the damage and loss data analysis carried out to define the stressor and to assign a logical value to the metrics within the row and columns of matrix.

4.4.1 Definition of the Stressor

The flash flood that occurred between the 11th and the 19th of October 2007 lasted for 8 days. Subsequent events such as the 2014 one, served to validate the metric assignment permitting to check where damages occurred again and again even though less severe.

4.4.2 Application of the Stressor and Evaluation of the Effects

The aim of this scenario is to identify the performance fallacies of the system in its current configuration considering the stresses that an event as the one occurred in October 2007 would cause. The event has been selected as reference, being the most damaging in terms of damage and losses, according to the analysis carried out on the compensation values provided by the CCS.

In order to apply the stress scenario on Benidorm, the event dynamics have been reconstructed. Specifically, an online research has been carried out to collect information regarding the impacts occurred due to the event of October 2007 (October 2007 Event Reconstruction Sources List[2]). Articles published online have been analysed to reconstruct the dynamics of the event.

[2] (i) https://www.informacion.es/benidorm/2007/09/01/intensas-precipitaciones-ocasionan-inundaciones-leves-7610389.html; (ii) http://www.alicantevivo.org/2007/10/jala-que-llueva.html; (iii) https://www.levante-emv.com/alacant/2007/10/13/crecida-rio-algar-obliga-desalojar-13524071.html; (iv) https://www.informacion.es/benidorm/2007/11/04/juez-aclarara-responsablidad-lluvias-colegio-7585997.html; https://histobenidorm.blogspot.com/2015/10/lasinundaciones-de-octubre-de-1971-en.html; (v) https://lamarinaplaza.com/2014/10/06/benidorm-actuara-en-orts-llorca-para-evitar-nuevas-inundaciones-pero-deja-en-el-aire-la-solucion-de-otros-puntos-conflictivos/; (vi) https://alicanteplaza.es/benidorm-sustituira-el-cruce-de-la-avenida-alfonso-puchades-por-una-rotonda; (vii) https://www.informacion.es/benidorm/2022/09/13/benidorm-gastara-4-3-millones-75384329.html.

The heavy rainfall of October 2007 left more than one hundred thousand users without telephone service throughout the Region of Marina Baixa where Benidorm is located. Monitoring system recorded more than 223 l/m^2 in 24 h in Benidorm. The intensity of the rainfall caused several damages in Benidorm. The Rincón area was the hardest hit due to the rains. The flooding of the Derramador and Barceló ravines and the large amount of water accumulated in the streets forced the Police to temporarily close roads such as Severo Ochoa and Ametlla del Mar, as in La Cala at the mouth of the Murtal ravine, Alfonso Puchades, Avenida del Mediterráneo. In addition, the heavy rain triggered a small landslide in the Los Almendros avenue. The Municipal administration declared that "the city is perfectly capable of coping with the rains" and that "despite the heavy rain, there have been no power outages and no recorded damage to infrastructures". Despite such claims, several vehicles were dragged by the water flow, basements, garages and shops were flooded in Rincón, such as also the Dynastic hotel, and tourists in La Torreta tourist complex had to relocate their caravans due to the damages occurred. Heavy damages were recorded in a private school (Lope de Vega International School), the water flow damaged the school's sports courts which were completely destroyed. The damages added up to more than 600,000 euros, and the sport club had to be provisionally relocated. A person had to be rescued by the Fire Department and the Local Police services after being trapped in a garage.

The collected information allowed to identify necessary metrics to consider for the resilience assessment. Following the analysis of the online resources, additional relevant information to understand the behaviour of the system, after the event (adopt phase) have been found.

In the following event of October 2014, at the confluence of Alfonso Puchades and Comunidad Valenciana avenues, in the La Torreta campsite problems were recorded again, in fact this area is one of the "confluence sections" every time it rains with a certain intensity, as is the area known as the 'Number One curve'. As has been the case for decades, most of the incidents are recorded in this area. Thus, the Local Police had to rescue three vehicles that were trapped in the area while the La Torreta campsite was "totally flooded" and the retaining walls had to be demolished to evacuate the accumulated water. It is the same place where the torrential rains of October 2007 already caused important problems. The private school Lope de Vega was affected again. Following this event, there were no major incidents except in the area indicated. The Municipal administration declared that the city went back to normality in couple of hours. The beaches have taken a little longer to recover, where the operators have been working "for 18 h".

The collection of the information has allowed to understand in broader terms the actions carried out, in terms of mitigation, considering the repeated impacts. After the 2014 event, the municipality has highlighted the necessity to implement some strategic actions, which have been pending for a long time, in the area of confluence of Alfonso Puchades and Comunidad Valenciana avenues, the so called "Curva del Number One" where repetitively water accumulates after torrential rain events. In order to prevent water accumulation in that area it would be necessary to divert all the water from the Imalsa area through Emilio Ortuño to the l'Aigüera ravine.

The work was estimated to cost about 5 million euros. In addition, another pending work was the channelling of a section of the Barceló ravine. In August 2021 the City Council of Benidorm has announced the investment of Hidrarqua (drinking water and sanitation services company) of 3.5 million euros for the renovation of infrastructures in Vía Emilio Ortuño as well as its surroundings. The project aims at intervening in the subsoil and in the urban scene to enhance both the drainage and the traffic management through the renewal of hydraulic infrastructures of the Drinking Water and Sanitation networks and the implementation of the rainwater network in Vía Emilio Ortuño and the construction of a roundabout on Alfonso Puchades avenue. The design contemplates the implementation of a new collector that will cross the subsoil of Avenida Alfonso Puchades and other surrounding streets, including different paving works with lateral catchments necessary to divert rainfall flows and avoid adverse effects that have already been partially solved with the previous work of the Esperanto street collector, in the Avenida del Mediterráneo. In September 2022, the tender for the project just started, according to the news, and the new estimation reached 4.3 million euro. In the area of interest there are first-residence homes, tourist apartments and hotels.

4.4.3 Resilience Assessment

As for the previous introduced Stress Testing, the scenario here considered has been outlined within the following Table 4.2.

By looking at the changes occurred in the city of Benidorm while observing the results of the damage and loss data analysis and the low performance of the system when the stressor is applied it has been possible to uncover planning choices and actions that have led to an increase of the flood risk in the municipality.

Water represents an emblematic factor for urban growth and development. Where it is scarce it represents a limiting resource, where it is abundant it represents a source of development.

For some activities, such as tourism, water is a driver of growth and expansion. Intense spatial urban and rural transformation occur around water bodies, leading to an excessive use of resources, in terms of both water and lands (Gössling and Peeters 2015). The south-eastern Spain coastal area is a concrete example of such dynamics. One of the most relevant natural risks in the Mediterranean area is flooding, specifically flash floods (Martínez-Ibarra 2017; Gaume et al. 2009). In the past decades it has been subject to a vast development of the housing sector and tourism called "sun and beach tourism model". Many flood prone areas have been occupied since the development of the tourism industry in the 1960s and channelling and other anthropic interventions have exploited river beds surfaces (Martínez-Ibarra 2017). Even though the area has benefited from investment in flood risk prevention and protection, economic losses caused by the impact of torrential rains are increasing (Gil-Guirado et al. 2014; Olcina Cantos et al. 2010).

Table 4.2 Metrics and score of stress testing B to implement within the resilience matrix

Stress testing B—evaluation of urban quality and resilience when applied the 2007 event								
Parameters considered		Performance score				Final score	Resilience matrix allocation	Reference s
		Very low [1]	Low [2]	Moderate [3]	High [4]			
Occupation of hazardous areas (first and second-home houses, touristic complex, infrastructures and leisure attractions)						1	Absorb/physical	PAM (2022)
Urban sprawl						1	Absorb/physical	PAM (2022), CCAP (2022)
Residential buildings constructed without measures to prevent damages						1	Prepare/physical	Chapter 5, Gaja (2008)
Second homes built with lower architectural quality during real estate boom (1998–2008)						1	Prepare/physical	Aznar-Crespo et al. (2020)
Labour immigrants for tourism employment						1	Absorb/economic	Aznar-Crespo et al. (2021)
Recorded losses	Commercial activities					1	Absorb/economic	Chapter 3

(continued)

Table 4.2 (continued)

	Residential sector					1	Absorb/economic	Chapter 3
	Offices					2	Absorb/economic	Chapter 3
	Vehicles					2	Absorb/economic	Chapter 3
	Industries					2	Absorb/economic	Chapter 3
Legal frameworks						1	Absorb/social	Chapter 3
Labour immigrants for tourism employment						1	Absorb/economic	Aznar-Crespo et al. (2022)
Workforce in tourism						1	Recover/economic	CCAP (2022)
Damages to private School, Hotel and Camping area (2007)						1	Absorb/economic	October 2007 event reconstruction sources

(continued)

Table 4.2 (continued)

Users without telephone service						2	Recover/social	October 2007 event reconstruction sources
Temporary closed roads						2	Recover/physical	October 2007 event reconstruction sources
Triggered landslide						1	Absorb/physical	October 2007 event reconstruction sources
Rescued people						4	Absorb/physical	October 2007 event reconstruction sources
Time to go back to normality for the city						4	Recover/physical	October 2007 event reconstruction sources
Beaches recovery time						3	Recover/physical	October 2007 event reconstruction sources
Transport infrastructure quality (secondary roads, motorways, airports)						4	Adapt/economic	Aznar-Crespo et al. (2022)
Direct damages in number one curve and of conflicting points						1	Absorb/physical	October 2007 event reconstruction sources

(continued)

Table 4.2 (continued)

Economic damages in number one curve and of conflicting points						1	Absorb/economic	Valencian cartographic institute
Flood Risk Map (PATRICOVA)						2	Prepare/informatio n	October 2007 event reconstructio n sources
Number one curve and of the conflicting points (2007, 2014)						1	Adapt/economic	CCS, October 2007 event reconstructio n sources
Damage and loss data set from different sources						3	Prepare/informatio n	October 2007 event reconstructio n sources
General Urban Development Plan (PGOU) dating to 1990						1	Prepare/social	Chapter 3
Vulnerabilities factors						1	Prepare/informatio n	Chapter 3
Mitigation and adaptation strategies implemented after the event						1	Adapt/social	October 2007 event reconstructio n sources

Pérez-Morales et al. (2018) in their study highlight the substantial increase in economic losses caused by floods, in the past years, in the coastal region of South-Eastern Spain where, Benidorm, is located. The main reason, resides in the effect of urban growth in flood-prone areas leading to the presence of river courses running in highly urbanized zones. In other words the cause of damages can be largely attributed to the ramping increase in the last 40 years of exposure, of assets and goods in flood prone areas. This increase in exposure along coastal regions has been higher in Southern Spain than in any other European country (Burriel 2008). Those performance dynamics are evaluated in the first part of Table 4.2, where a very low score is assigned to metrics. Aznar-Crespo et al. (2020) define tourist destinations in the province of Alicante as a model of social vulnerability creation for tourism businesses, tourists themselves and local population. Buildings in the Alicante coast have been built with poor architectural quality during the real estate boom (1998–2008) causing intense environment degradation.

Legal frameworks have been evaluated with a score equal to 1 (very low performance score) since they are accountable for the socio-economic growth and the rise of housing construction occurred due to the intense real estate boom. This has been the result of a process that began in the Fifties, urban regulations promoted its acceleration (Gaja 2008) and led to an evident increase of vulnerability and exposure of both people and assets due to poor management of flood-prone areas by government institutions and questionable regulations that have been adopted (as discussed in Chap. 3).

While considering the differentiation of the recorded losses of the CCS dataset, after the 2007 event, a different score has been assigned to the different categories. According to the damage and loss analysis presented in Table 4.2 commercial activities and residential buildings have suffered the highest damage followed by offices, vehicles and industry. The performance score is assigned accordingly equal to 1 for the first two most damaged categories and equal to 2 for the rest, considering the developments occurred in the past years in the fast-growing mass tourism complex system analysed. The assessment is allocated in the adsorb/economic intersection cell of the resilience matrix.

The analysis carried out and used to implement stress testing B showed that both the flood event of 2007 taken as a reference, and the subsequent one occurred in 2014 have caused significant damages in the same area, which has been later defined as "Number One Curve" being a recurrent confluence point and to the same assets and streets (Fig. 4.4) which has been evaluated in the process in terms of physical damage and economic losses with a very low score, given the repetitive occurrence of the damages in the years, which might occur again if at the present an event with similar features would occur, since no mitigation measure has been implemented yet (considering that no information have been found online on the execution and completion of the project of Hidrarqua—see Sect. 4.4.2—which aims at implementing some structural mitigation measures in the area). The two performance evaluations have been respectively allocated in adsorb/physical and adsorb/economic intersection of the resilience matrix given that up to now the system shows a clear low adaptive capacity. The event of 2007 caused extensive damage to the Dinastic Hotel, La

(a)

(b)

(a)

Fig. 4.3 Impacts of the 2014 flood in Av. Alfonso Puchades (**a**) and Av. Mediterráneo (**b**) and (*Source* PAM 2022)

Torreta camping and the to the International School Lope La Vega; those last two have been impacted in 2014 again, such as the same roads (see Fig. 4.3); a very low performance score, equal to 1, has been assigned to this metric considering that no improvement or change have actually occurred up to now.

The following map (Fig. 4.4) allows to understand the location in the city of such confluence sections where damages have occurred both in 2007 and 2014. The map has been reconstructed by localizing the points described in the analysed sources in order to have a spatial allocation of the information. Clearly the occurrence of the same damages in the same areas after seven years from one of the most damaging events, and the fact that no improvement, in terms of structural or non-structural mitigation measures, has been recorded, denotes a very low adaptation capacity of the system, which wasn't able to learn from past to enhance its current configuration. A very low performance of the system in terms of resilience and high vulnerability up to today, till the project will be completed, is considered. Such behaviour is evaluated through the Number One Curve and of the confluence Sections (2007, 2014) metric and allocated in the adapt/economic cell of the resilience matrix.

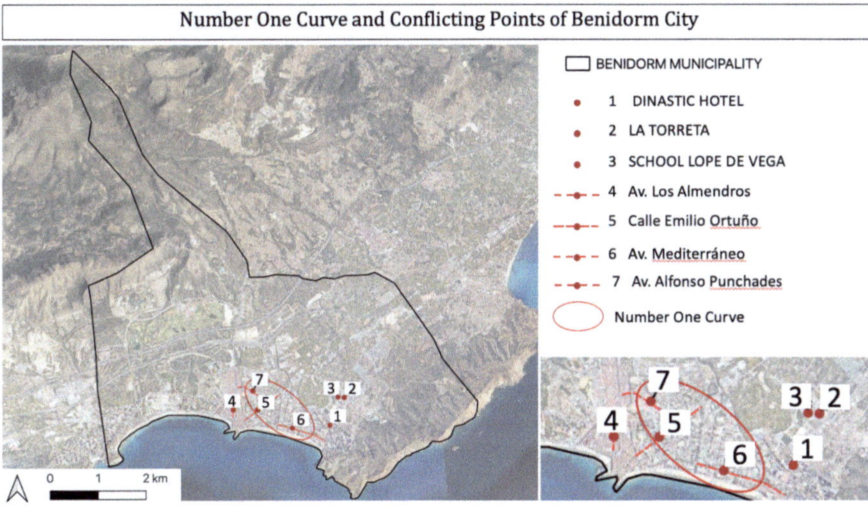

Fig. 4.4 Map of the number one curve and of the confluence sections for the events 2007 and 2014

The following Fig. 4.5 combines the different analysis carried out (i) damage and loss data distribution, (ii) repeated punctual damages in the Number One Curve and the nearby assets together with the Flood Risk Map of the PATICOVA. The map shows that:

- no areas at risk are present in the 03500 postal code (northern area), which has been the one in which most of economic losses occurred both in terms of the whole

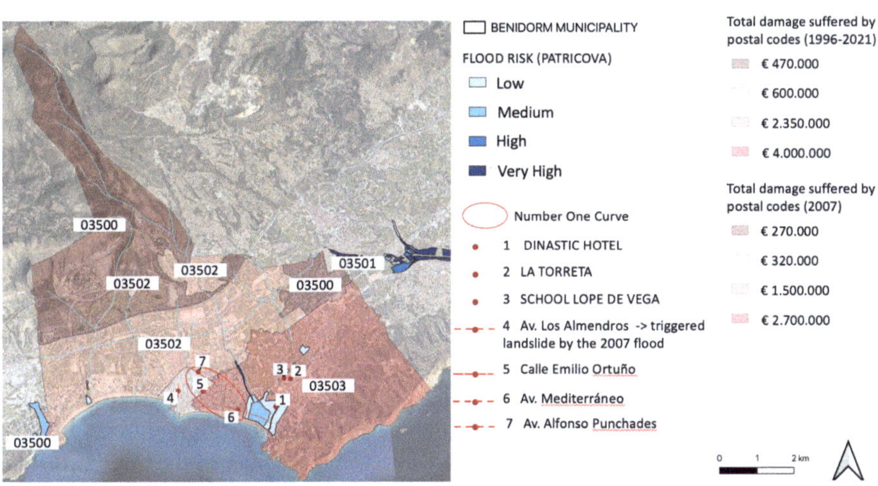

Fig. 4.5 Map representing the most damaged areas, critical points impacted by the floods from 2007 and 2014 and the flood risk areas of the PATRICOVA

time range analysed (1996–2021) and for the most damaging event occurred in 2007;

- there is no one-to-one correspondence between the points highlighted in Fig. 4.4 (hotel, school, camping area and Number One Curve) and the areas at risk represented by the PATRICOVA but only a partial overlap.

Therefore, it becomes evident that the PATRICOVA elaboration is a hazard assessment rather than a risk one. PATRICOVA lacks a series of data in the elaboration (a metric already considered in stress testing A), such as exposure, that could bring to a more meaningful representation of the real areas at risk. Proper analysis at local level would provide the necessary detail to address a resilient urban planning, which can act as a prevention measure. For this reason, the Flood risk Map (PATRICOVA) has been evaluated with a low score since taking it as a reference conveyed changes, up to today, which increased exposure and vulnerability, in the urban developments realized after the its introduction. The score assigned is equal to two in the implementation of the stress testing and it is allocated in the intersection between the prepare phase and the information domain of the final resilience matrix.

As a proof of what has been stated in the present research, damage and loss data constitute a key knowledge element to delineate damage dynamics on the territory and highlight critical areas that deserve attention in order to reverse the negative system performance and achieve resilient behaviour. Even thought, the results of the analysis carried out at the postal code level present the clear limit of the areal distribution, which doesn't allow to have point shaped data but rather values spread across the whole postal code area. As a consequence, between one of the last rows of Table 4.2 the damage and loss data set from different sources metric has been evaluated with a score equal to 3 meaning moderate performance allocated in the prepare/information cell intersection of the resilience matrix, given the knowledge that has been retrieved from them.

The analysis of the Municipal General Urban Development Plan of Benidorm (PGMO), currently in force, allowed to assign the described and the following performance score. The plan dates back to 1990 and it has never been updated, apart the addition of 22 localized modifications approved over the years.

Since it has not been possible to retrieve any digital or vectorial file of the PGMO, the following map (Fig. 4.6) has been produced by tracing and georeferencing the original scanned paper map, downloaded from the official site of the municipality, using the GIS software in order to ease the analysis process.

Once obtained the overview of the initial idea of the masterplan, which already contained classifications that moved against the compact urban model implemented in the initial 1956 masterplan (see Fig. 3.5), becomes key to observe the current land use. The land use cover data used to produce the map on the left side of Fig. 4.7, have only been made public very recently on the website of the Cartographic Institute of Valencia. Figure 4.7 allows to compare the initial master plan of the city according to the PGMO of 1990 (right side) and the city development over the last decades until today (left side).

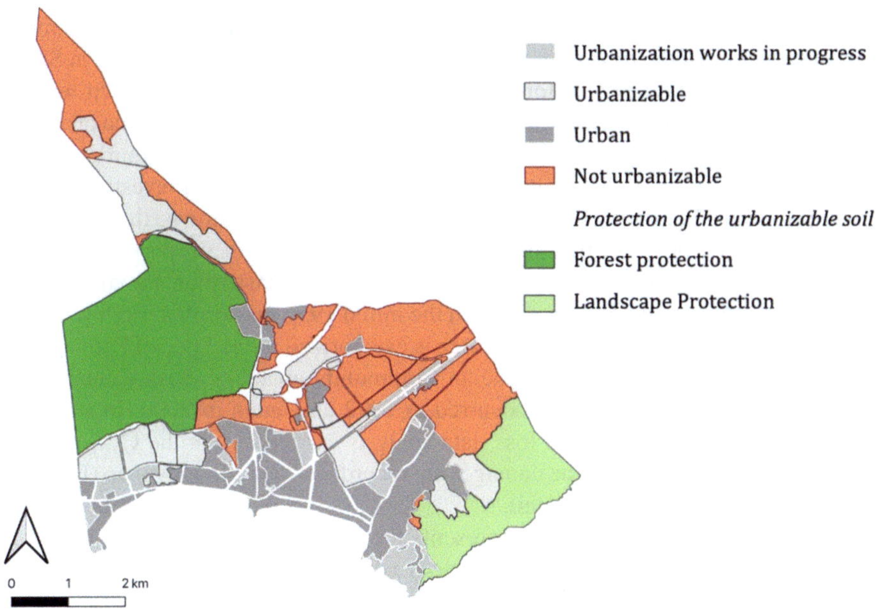

Fig. 4.6 Vector elaboration of the general urban development plan, 1990

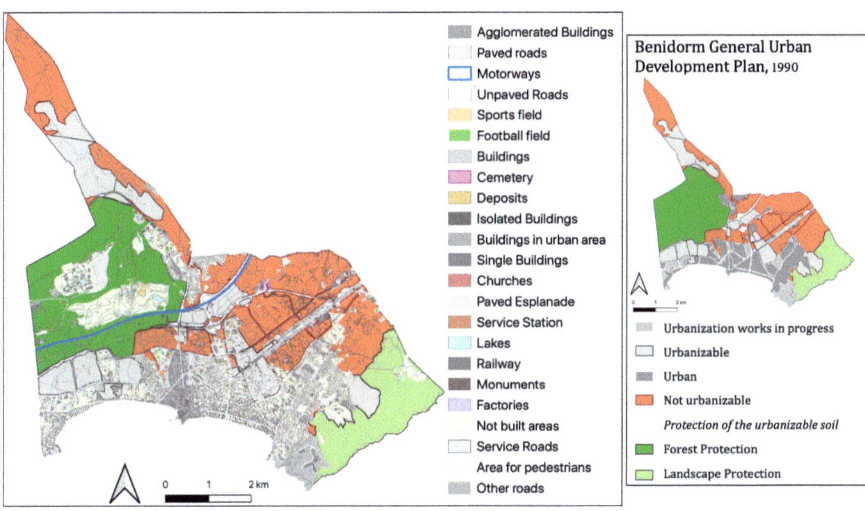

Fig. 4.7 Current land use of Benidorm city

Recalling the analysis carried out in Sect. 3.2, controversial decisions were made as a consequence of the forest fire that in 1992 devastated large land surface in Sierra Cortina, the area classified as forest protection in the PGMO and known as "the largest pine forest in the Mediterranean", where nowadays the Terra Mítica project is located. The masterplan of 1990 was an instrument needed to increase tourist quality standards (I Baidal et al. 2013), it facilitated private developers in investing in new leisure proposals. Following the forest fire, public investments allowed the realization of the Terra Mítica project which involved the construction of theme parks, golf fields and other leisure attractions and modifications to the 1990 masterplan, moving against the compact urban model implemented in the initial 1956 masterplan. Following the MTP (2019), the Sierra Cortina forest area is categorized in the Special Plan against the Risk of Forest Fires of the Valencian Community as a High Risk Zone, where most significant forest fires have occurred in 1992 and again in 2009. In addition, following the flood damage and loss data analysis carried out on the CCS dataset, that is the area where most of the damages are recorded. The following map (Fig. 4.8) shows the described area into detail.

This case is a key point which represents the exact proof that if urban planning is not synergically accompanied by knowledge deriving from disaster risk reduction and resilience approaches it can lead to conditions of increased risk for the area under development. Acting on the territory without having a longer term perspective which takes into account systemically all its features and functions increases risks and even generates new ones, in this case specifically by increasing not only exposure but most importantly vulnerability.

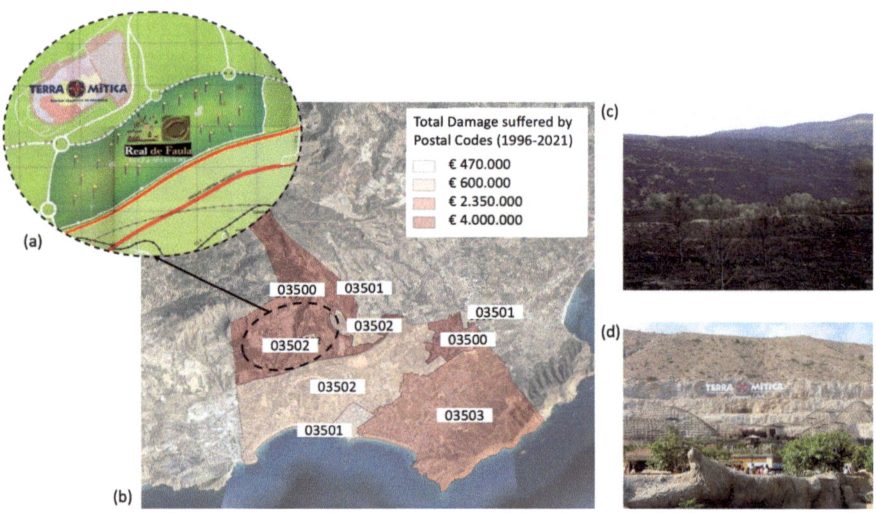

Fig. 4.8 Area of the the Terra Mítica project (*Sources* **a** Comunitat Valenciana, **b** elaboration of the author, **c**, **d** Asociación Cultural Alicante Vivo)

The changes made to the natural configuration of the area of the Sierra Cortina, first due to the forest fire (i.e. loss of the tree vegetation) and then to the construction activities in the whole park area (impact on the geomorphology) increased its vulnerability (i) to floods as shown by the analysis of the damage and loss data of the CCS, (ii) to forest fire due to the significant change in land use; where previously there were fields of crops, now they are large urbanizations scattered with chalets and bungalows, which have generated large areas of urban-forest interfaces in the vicinity of the Sierra Cortina (as also for the Natural Park of Serra Gelada) where the risk of forest fire is very high (CCAP 2022). This example also clearly explains that knowledge past events is fundamental as relying on hazard maps only, as in this case, without accounting for vulnerabilities factors provides a very partial and even wrong perception of the risk.

The final results are represented in the following Resilience Matrix (Fig. 4.9), where the colours from red to green represent a gradually a more resilient behaviour of the system when stressed across its domain and between the prepare-adsorb-recover-adapt phases.

The proposed methodology allows to highlight the problems and criticalities of the system analysed. Areas on which planning should act to enhance the current status of the city become clear. Stress Testing B considers the current configuration of the city and makes reference to the event of 2007 for the recorded damages and the impacts experienced on the different functions of the city. Given increasing exposure and vulnerability trends, in the absence of structural and non-structural proper mitigation measures, or until the latter will not be completed, with a similar type event, the same damages and losses recorded in past, if not larger, should be expected.

This is clearly represented within the absorb/social cell of the resilience matrix, where the prepare/physical and adsorb/physical assessment is mainly given by the

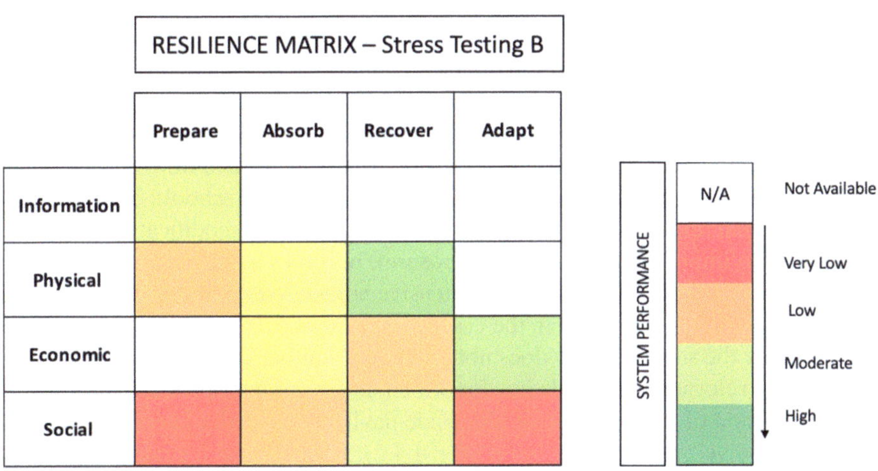

Fig. 4.9 Resulting resilience matrix of stress testing B

exposure and vulnerability of the system increased along the years due to the questionable legal urban development frameworks adopted at national level and locally. The adsorb phase low result are given by the high values recorded of the damage and losses due to the event. While a better performance can be seen on the right side of the matrix, within the recover and adapt phases. Such result in the "recover" phase is linked to the analysis of the post-event phase carried out on the online sources, from which it has been possible to reconstruct that the recovery occurred smoothly allowing to re-establish the initial conditions of the system. The adapt/economic cell presents good results due to the assessment of the transport infrastructure quality (secondary roads, motorways, airports) which are usually well developed in those area of mass-tourism attraction. Lower values are present in the social domain given that after the event no mitigation and adaptation strategies have been implemented. Structural mitigation measures have been designed, but it is only in August 2021 that the City Council of Benidorm has announced the investment of a drinking water and sanitation services company of 3.5 million euros for the renovation of infrastructures in the Number One Curve and the confluence area.

4.4.4 Use of Stress Testing to Support Planning

The Municipal General Urban Development Plan precedes the new regulations on flood risk (PATRICOVA) and, as showed from Figs. 4.5 and 4.7 analysed jointly, fails to consider the damage dynamics of the area. In this case our proposed methodology would allow to uncover the critical areas of the system that need improvements and to tailor specific measures to undertake in order to guarantee the achievement of urban resilience while contemplating objectives of risk prevention and mitigation.

For instance, while looking at the specific area of the Number One Curve and the adjacent assets which have been repetitively damaged by flood events (Fig. 4.5), it would be beneficial to propose as a preventive planning strategy the "Functional Urban Redefinition" (Ridefinizione Funzionale Urbana in Italian), proposed by Gazzola (2019), which on the basis of principles of urban flexibility, as written by her, proposes to reduce urban exposure through changes of use between different urban functions, namely school education and leisure which are the activities carried out in the flood risk prone area (point 1 hotel, 2 camping, 3 private school). This strategy, similar the areal relocation but based on a different approach (modification of the use of the asset to improve the functional exposure), becomes a preventive intervention tool in already urbanized areas where extensive adjustments and improvements are not always possible, exactly as in the case study analysed.

Obviously the same strategy does not fit for the Number One Curve mainly characterized by relevant crossroads but for which structural measures are foreseen as the construction of a new collector alongside paving works with lateral catchments necessary to divert rainfall flows (see Sect. 4.4.2). However, such strategy could not fit for the Terra Mítica (see Fig. 4.8). considering the type of activities carried out in the park as well as its dimensions and the additional soil consumption that would is expected there. Mitigation measures aimed at reducing both exposure and vulnerability should be foreseen in the urban plan in order to enhance the system behaviour and its resilience.

4.5 Stress Testing C: Applying the Climate Change Impacts and Urban Growth Projections

By making reference to the Municipal Action Plan and the CCAP (2022), Stress testing C considers the impacts of climate change on the territorial system in its whole, considering the tourism, infrastructures and residential sectors, hypothesizing no changes on the current configuration.

4.5.1 *Definition of the Threat*

Stress Testing C gets shaped on a scenario which takes into account the results of the climate model for the city of Benidorm as a stressor to test its resilience under its effect. This operation does not consider the concretization of the 13 lines of actions proposed by the CCAP (2022) described in Sect. 3.3.1. Such a Scenario would serve to highlight the disruptions that the system could undergo if no mitigation and adaptation measure is implemented.

4.5.2 *Application of the Stressor and Evaluation of the Effect*

In order to evaluate the system performance once the stressor is applied the following impacts defined in the CCAP (2022) are considered:

- Increase in average temperatures, both minimum and maximum (increasing number of tropical nights which can generate a considerable increase in water stress for the vegetation);
- The number of precipitations will be reduced: concentrated in fewer episodes with less abundant rain in spring and more abundant in autumn, but with hourly intensification of the rains (torrentiality) with increased runoff;
- Rise in the number and duration of droughts and intensification of their effects, leading to an increase in water stress of the vegetation, increasing the risk of (i) forest fires, (ii) erosion, and (iii) desertification;
- Less marked seasonality;
- Increase in weather extremes, episodes of torrential rains, severe droughts, heat waves, cold waves, and maritime storms.

Following the risk assessment carried out in the CCAP (2022) the economy, the community and the environment within the municipality of Benidorm are sensibly linked to the impact of extreme heat, droughts, fires and floods, some of which can produce synergies between them increasing risk levels, such as for instance heat waves, droughts which can enhance forest fire risk; in addition to the increase in evaporation and evapotranspiration due to the higher average temperatures which worsen the arid conditions of the soils boosting the erosion process.

4.5.3 Resilience Assessment

The scenario here considered has been outlined within the following Table 4.3.

Foreseen risks and impacts due to climate change as modelled within the plan are considered with adaptation measure is implemented, thus considering that the short time of implementation and the little monetary investments planned for the line of action proposed in the CCAP (2022) do not seem to be feasible. The metric is therefore considered and the performance score applied is equal to 1.

A low performance is assigned also to the urban environment as the latest Masterplan, dated to 1990, precedes the new regulations on flood risk (PATRICOVA). The masterplan has never been updated. Therefore, the urbanized areas/zooning has never been revised according to the hazardousness of the area according to the PATRICOVA regulations. In addition, as the PATRICOVA does not contemplate any consideration about climate change impacts, then this constitutes a further aspect of low performance of the system to be considered in this stress testing.

The absence of local plans, of risk assessments of the critical infrastructures, of vulnerability studies assessing the presence of touristic activities, of tourists themselves and the aging population (both of tourists and residents) constitutes a significant limitation to preparedness in case of extreme events.

Another relevant factor is related to the poor performance of the system in terms of low economic diversity, being the one of Benidorm a mainly tourism-dependent economy therefore a more vulnerable one especially in the face of the impacts derived from the models of Climate Change. The main economic engine in Benidorm is tourism, 94% of the people affiliated with Social Security in 2020 work in the tourism service sector and 89% of the companies in the city work in the tourism sector. Of those people 40% work in the hotel industry, followed by those who work in wholesale and retail trade. The diversification of the economy is evaluated with a very low score allocated in the adapt/economic intersection of the resilience matrix.

Another important matter in such a context noteworthy of attention, resides in the social vulnerability to natural hazards of tourist destinations. The constantly changing population, due to the cohabitation of space by visitors, tourists and residents in the exposed city creates important administrative and practical difficulties. Massive flows of foreign tourists and residents of all ages, temporarily or seasonally visiting the city, with seasonal population often exceeding residents, is a condition that is not properly analysed nor considered in any of the official documents. This fluctuation, with

Table 4.3 Metrics and score of stress testing C to implement within the resilience matrix

Stress testing C—applying the climate change impacts and urban growth projections							
Parameters considered	Performance score				Final score	Resilience Matrix Allocation	References
	Very low [1]	Low [2]	Moderate [3]	High [4]			
Municipal action plan for the risk of floods					3	Prepare/information	PAM (2022)
Climate change adaptation plan					3	Prepare/information	CCAP (2022)
Climate change adaptation plan lines of action					1	Prepare/economic	CCAP (2022)
General Urban Development Plan (PGOU) dating to 1990					1	Prepare/social	Chapter 3
Flood Risk Map (PATRICOVA)					1	Prepare/information	Chapter 3
Increase of urbanized area					1	Absorb/physical	CCAP (2022)
Population increase					1	Absorb/physical	CCAP (2022)
Inhabitants/day in high season					1	Absorb/physical	CCAP (2022)
Road access in residential areas with high building density for emergency and rescue services					2	Absorb/physical	Aznar-Crespo et al. (2020)
Occupation of flood prone areas					1	Absorb/physical	CCAP (2022)
Modified the environmental configuration increasing the exposure to forest fires					1	Adapt/physical	CCAP (2022)
Foreigners inhabitants (28.5% of the total population)					3	Absorb/physical	CCAP (2022), Aznar-Crespo et al. (2020)
Age of residents - Population is between 30 and 65 years old					2	Absorb/social	CCAP (2022)

(continued)

Table 4.3 (continued)

Aging index with an increasing trend (arrival of elderly foreign population)						2	Recover/physical	CCAP (2022)
Diversification of the economy, tourism-dependent economy						1	Adapt/economic	CCAP (2022)
Water demand						3	Adapt/social	CCAP (2022)
Foreseen criticalities	Increase in average temperatures					1	Recover/physical	CCAP (2022)
	Rains intensification					1	Adapt/physical	CCAP (2022)
	Rise in number and duration of droughts and intensification of their effects					1	Adapt/physical	CCAP (2022)
	Less marked seasonality					1	Adapt/physical	CCAP (2022)
	Increase in weather extremes, episodes of torrential rains, severe droughts, heat waves, cold waves, and maritime storms					1	Adapt/physical	CCAP (2022)
Foreseen impacts	Torrential rains and associated runoff can generate an increase in waste to be managed					1	Recover/physical	CCAP (2022)
	Extreme weather events will affect transit routes					1	Recover/physical	CCAP (2022)
	Increased risk of slope instability					2	Absorb/physical	CCAP (2022)
	Increase in maritime storms and the rise in sea level will favour the deterioration of the infrastructures					1	Recover/physical	CCAP (2022)
	Increased exposure of people and assets, torrential rains, strong winds, storms					1	Recover/physical	CCAP (2022)

(continued)

Table 4.3 (continued)

Foreseen impacts	Increase in extreme meteorological phenomena, maritime storms and rise of sea level, can have a direct impact on tourists and infrastructures (promenades, beaches, marinas)					1	Recover/social	CCAP (2022)
	Loss and regression of the beaches					1	Adapt/physical	CCAP (2022)
Implementation of a local plan for flood prevention						3	Adapt/physical	PAM (2022)
Risk assessment of the critical infrastructures						1	Absorb/information	Chapter 3
Risk maps and cartography						2	Adapt/information	Chapter 3
Take into account climate change projections in the development of any new plan or programme						3	Recover/social	Chapter 3
Vulnerability studies						1	Adapt/social	Chapter 3
Linguistic barriers of tourists and foreigners						1	Adapt/social	Aznar-Crespo et al. (2020)

periods of high demographic density, and the lack of risk knowledge and awareness by tourists and lack of familiarity with the local environment, require appropriate emergency management. Aznar-Crespo et al. (2020) define tourist destinations in the province of Alicante as a model of social vulnerability creation for tourism businesses, tourists themselves and local population. Essential features of those areas highlighted in the study, and considered within the performed stress testing, are related to the scarce awareness of the environment including of potential natural hazards and measures to undertake for protection of the tourists and lack of integration of short-stay tourists, A further vulnerability factor is constituted by the occupation of second-homes in isolated hazardous areas within the Benidorm agglomeration and camping areas located in zones that have been repetitively flooded in the years. Finally, the reduced availability of open space for public use, viability problems due to high building density and extraordinary population density during the high season make response following the impact of an extreme particularly challenging.

However, to balance "negative" indicators of increased vulnerability as defined by Aznar-Crespo et al. (2020), the development of such "sun and beach" areas contributes to some extent also to the opportunities for resilience, in terms of large development of urban lifelines included enhanced transportation main line for the tourists to reach the final destination, as evident in Benidorm city.

However, the urbanization phenomenon, as reported in the Climate Change Action Plan (2022), has led to an increase of both exposure and vulnerability to the hazards threatening the area, considering that the urban growth took place in flood prone hazard areas and has modified the environmental configuration increasing the exposure to forest fires.

The lack of risk management good practices, no seismic micro-zonation (Aznar-Crespo et al. 2020), no multi-hazard assessments and lack of risk mapping including besides hazards analysis also social, physical, and systemic vulnerability considerations have not permitted to identify comprehensive enough disaster risk mitigation and climate change adaptation measures.

Last but not least, water resources in Benidorm represent a critical element due to both meteorological and climatic conditions, with increasingly irregular rainfall and periods of drought, and to the demand side that has been growing substantially over the years. According to the Climate Change Action Plan (2022) the Municipality has worked to achieve self-sufficiency, due to the scarcity problems faced in the past. Currently, Benidorm has an efficient drinking water supply and distribution system, with insignificant losses of water in the network due to the urban configuration characterized by dense urban area developed following a vertical design.

The urban growth experienced from the 1960, has led to a change in traditional land uses favouring the occupation of hazard prone areas, increasing exposure and vulnerability. When Figs. 4.5 and 4.7 are analysed jointly it becomes evident that the areas which were not considered developable in the initial master plan of 1990 are the ones where most of the damages are recorded nowadays, specifically 03500 postal code which was an area of forest protection and part of the 03503 postal code which initially categorized as landscape protection area. A result that did not reflect the initial expectations, considering that following the preliminary analysis of the

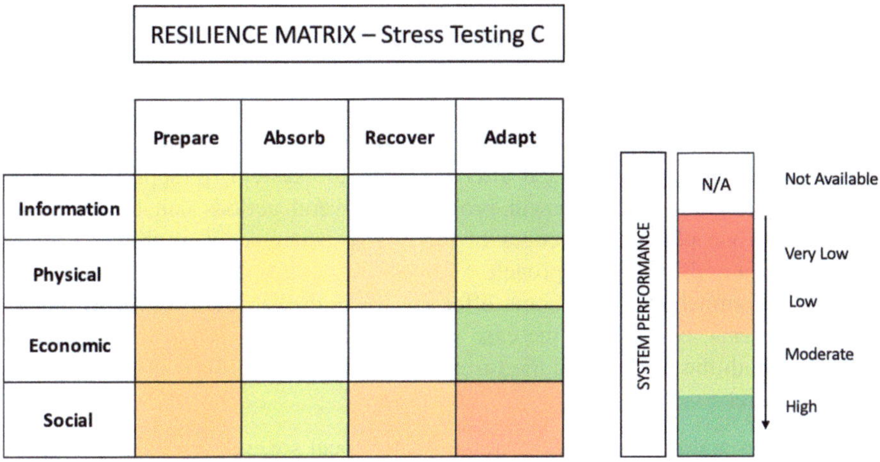

Fig. 4.10 Resilience matrix of stress testing C

city, a high level of urban exposure has been identified in the southern part of the city, in proximity of the coastal area (Ponente and Levante Beaches), proving that vulnerability plays an important role in the overall risk and resilience levels of a system. However, considering the impacts of climate change, as highlighted in the CCAP (2022), the increase in maritime storms and the possible rising sea levels can generate major problems in the infrastructure of the first coastal strip, in addition to the possible environmental damage, in the first coastline and the Serra Gelada Marine Reserve.

The following Fig. (4.10) reports the results obtained by applying the adverse scenario of stress testing C considering climate change impacts.

Stress Testing C considers the climate change projections developed within the municipal plans, the prepare/information cell of the matrix shows a performance tending towards a moderate score given the commitment of the municipality in getting aligned with the challenges and needs posed by current and future risks. However the available plans and document have been evaluated and the general behaviour of the system can be classified between low and moderate performance level. If no inversion of the current system dynamics would occur, a low performance would be experienced in adsorption, recovery phases when the system is impacted by stresses of climate change.

4.5.4 Use of Stress Testing to Support Planning

Some strategic planning alternatives to enhance urban resilience in Benidorm can be proposed while envisaging risk prevention and mitigation objectives in the face of climate change foreseen impacts. The lack of integration between territorial governance and risk mitigation and prevention processes, as well as resilience building, becomes evident in the planning approaches that have been held over time in the city of Benidorm; and it manifests in the criticalities uncovered during the research, finding an explanation in the actions that have been performed on the territory.

Following the results obtained after stressing the system by applying climate change impacts and urban growth projections several actions can be envisaged to support urban planning practices of the city to better adapt to climate change according to a multi-hazard approach.

Some recommendations become effective for both the stress testing B and C. In general terms, considering the case study peculiarities, suitable tools aimed at enhancing both the current and the future configuration of the city (i.e. the system behaviour) are:

i. mapping and detailed studies developed at local scale for all hazards, taking into account exposure and vulnerability;
ii. adoption of specific technical standards, for instance enforcing building structural standards such as pilotis buildings in flood prone areas making sure though that they are also seismically resistant;
iii. abatement ordinances for facilities in dangerous areas;
iv. real estate disclosure obliging contractors disclose risk potential to buyers or creating specific clauses in contracts for the sale and rental of properties at risk especially considering the scarce knowledge of the environment of the tourists which represent a large portion of the population;
v. implementation of standards and regulations for new development areas taking into account knowledge such as the one that the proposed methodology conveyed and that enhanced mapping and detailed studies proposed in point (i) could produce;
vi. prohibition of further development in the most hazardous zones as identified by our assessment, following the revision of currently available hazard maps including projections of climate change;
vii. integration of structural measures that are on the way to be constructed within the urban plan, following a new dynamic modelling of the hydraulic conditions of the city to determine areas that should be subject to relocation even when the structural measures will be accomplished. The enhancement of the link between emergency plans for the city and urban plans (Menoni 2020) represents another crucial strategy to implement.

Finally nature-based solutions can be tailored on the results obtained, therefore considering properly the insights emerged from the damage and loss analysis such as the vulnerabilities of the system, to enhance resilience while mitigating and setting the ground for adaptation, for current risks and potential ones foreseen due to the impacts of climate change. Relocation or functional relocation, as proposed in stress testing B, of the most vulnerable functions are other type of functional mitigation measure for the system analysed.

References

P. Aznar-Crespo, A. Aledo, J. Melgarejo-Moreno, Social vulnerability to natural hazards in tourist destinations of developed regions. Sci. Total Environ. **709**, 135870 (2020)

CCAP, in *Climate Change Adaptation Plan* (2022). Available at: https://contenidos.benidorm.org/sites/default/files/descargas/2022-05/PLAN%20DE%20ADAPTACIÓN%20ANTE%20EL%20CAMBIO%20CLIMÁTICO%20DE%20BENIDORM_OCT21_IC.pdf

S.L. Cutter, C.G. Burton, C.T. Emrich, Disaster resilience indicators for benchmarking baseline conditions. J. Homeland Secur. Emerg. Manag. **7**(1) (2010)

T. De Groeve, K. Poljansek, D. Ehrlich, in *Recording Disaster Losses: Recommendation for a European approach* (EUR 26111 ENPublications Office of the European Union, Luxembourg, JRC83743, 2013)

B. De Orueta, L. Eugenio, La "década prodigiosa" del urbanismo español (1997–2006). Scripta Nova **12**(270), 64 (2008)

A. Faiella, Agriculture damage data collection: a model for reconstructing comprehensive damage dynamics. Progr. Disast. Sci. **7**, 100112 (2020)

A. Faiella, S. Menoni, M.P. Boni, M. Panoutsopoulou, T. Thoma, S. Salari, N. Rueda, Enabling knowledge through structured disaster damage & loss data management system. Sustainability **14**(10), 6187 (2022)

F. Gaja, in El "tsunami urbanizador" en el litoral mediterráneo, in *El ciclo de hiperproducción inmobiliaria 1996–2006* (Scripta Nova, 2008), p. 12. Available at: http://www.ub.edu/geocrit/sn/sn-270/sn-270-66.htm

E. Gaume, V. Bain, P. Bernardara, O. Newinger, M. Barbuc, A. Bateman, ... A. Viglione, A compilation of data on European flash floods. J. Hydrol. **367**(1–2), 70–78 (2009)

V. Gazzola, in *Uso del suolo e riduzione del rischio naturale in ambito urbano* (Maggioli Editore, 2019)

S. Gil-Guirado, A. Pérez, F. Lopez, M. Barriendos, Review of flood alert thresholds based on the socioeconomic impact of catastrophic events in the southeastern Spain, in *EMS Annual Meeting Abstracts* (vol. 11, 2014)

S. Gössling, P. Peeters, Assessing tourism's global environmental impact 1900–2050. J. Sustain. Tour. **23**(5), 639–659 (2015)

I. Baidal, J.A.I. Sánchez, J.F.V. Rebollo, The evolution of mass tourism destinations: new approaches beyond deterministic models in Benidorm (Spain). Tour. Manag. **34**, 184–195 (2013)

E. Martínez-Ibarra, Geographical analysis of flash floods in tourism areas in the north of the province of Alicante. Papeles de Geografia, **63**(Special Issue), 65–81 (2017)

S. Menoni, Risk Mitigation and Resilience of Human Settlement. In Balducci, A., Chiffi, D., & Curci, F. (Eds.). Risk and Resilience: Socio-Spatial and Environmental Challenges (pp. 27–47). Springer International Publishing (2020)

MTP, *Plan Territorial frente a Emergencias Municipio de Benidorm* (2019). Available at: https://contenidos.benidorm.org/sites/default/files/documentos/ptme_benidorm_2019lpd.pdf

J. Olcina Cantos, M. Hernández Hernández, A.M. Rico Amorós, E. Martínez Ibarra, Increased risk of flooding on the coast of Alicante (Region of Valencia, Spain). Nat. Hazard. **10**(11), 2229–2234 (2010)

PAM, *Municipal Action Plan Against The Risk Of Floods In Benidorm* (2022). Available at: https://contenidos.benidorm.org/sites/default/files/descargas/2022-03/1_pdfsam_pam_inundaciones_benidorm_14022022_lpd.pdf

A. Pérez-Morales, S. Gil-Guirado, J. Olcina-Cantos, Housing bubbles and the increase of flood exposure. failures in flood risk management on the Spanish south-eastern coast (1975–2013). J. Flood Risk Manag. **11**, S302–S313 (2018)

Chapter 5
Discussion and Operationalization of the Results

Abstract This chapter presents a discussion of the results and an evaluation of the proposed methodology providing a balanced view of potential advantages and implications related to its application. Then a synthetization of the procedure developed is proposed in order to streamline the process in a form of guidelines so that it can be applied flexibly and pragmatically in future works.

5.1 Discussion

In this chapter the application of the proposed methodology to the case study of Benidorm will be discussed with a view to show what can be gained in terms of knowledge from the analysis on the specific context and if and at what conditions it can be abstracted and used elsewhere.

The initial analysis carried out revealed that even though the relationship between the two practices of spatial planning and disaster risk reduction is at the core of numerous policies and the awareness of the connection between the two has been evolving in time, their concrete integration is still far in practice. Literature and policy landscape analysis supports the idea to complement traditional risk-based approaches with resilience assessment and management. Resilience becomes the guiding principle to address complexities related to climate change adaptation and disaster risk reduction in cities, which are among the most complex systems that humans have created, due to the dynamic interdependence of people, services, economic activities and infrastructures densely concentrated in space.

Being resilience intended as a process which learns from past experiences making use of a set of different skills and resources, empirical evidence of the impact that a disaster has on a territory, is an opportunity for the enrichment of the base-knowledge needed to implement a proactive and adaptive spatial planning which integrates concepts of disaster risk reduction framed within the concept of resilience. However, research shows that there is still room for improvement in current practices of damage and loss data collection. Often available data are not reliable as they cannot be grounded on a proper collection procedure, their management and coordination is fragmented among different authorities and organizations that perform the collection

A. Faiella, *Learning from Disasters to Enhance Resilience of Urban Complex Systems*, PoliMI SpringerBriefs, https://doi.org/10.1007/978-3-032-07356-3_5

for different purposes following their own internal approaches. Those data are rarely exhaustive as either they are too aggregated or they are hazard- and sector-specific. In addition they do not cover the entire time frame from initial detection to proper assessment and eligibility decisions.

The novel methodology here proposed, designed to overcome the shortcomings of available approaches, is developed to be applied on the system dimension, which in this case is a city, taking resilience into full account. The stress testing and the resilience matrix, at the core of the methodology, are applied to the selected case study of this work, Benidorm city.

The urban system of Benidorm has been analysed into detail. The findings of such an extensive analysis allowed to define the final scores assigned to stress testing performed and organized in the resilience matrix. Benidorm is a fast-growing mass tourism-specialized coastal city in hazard prone area. Several hazards threaten the city, ranging from landslides, forest fires, droughts to urban floods. The investigation of the development of the city over time revealed that past choices led to an increase exposure and vulnerability, as revealed also by analysing damage and loss data provided by the CCS. Those findings become essential for the metric evaluation of the stress testing once applied to Benidorm. As a final result of the application three resilience matrixes where the stress testing outcomes are organized, deliver a resilience assessment of the city. The three stress tests have been handled in a separated fashion and results have not been combined in a unique resilience matrix as they refer to different temporal scales. The three separated results allow to have a resilience monitoring over time, i.e. a comprehensive comparison between the resilience assessment of the different temporal phases of the system considered (present, past and future).

Embedding damage and loss data in the stress testing has been particularly useful to propose strategic planning alternatives to enhance urban resilience in Benidorm. The damage and loss data analysis revealed that the year that has recorded the highest damages is 2007, followed by 1997 and 2009. The spatial analysis of recorded data shows that the area that received more pay-outs, in the time range analysed which goes from 1996 to 2021, is the 03500 postal code area with almost 4 MLN EUR compensated damages, followed by 03503 area with more than 2 MLN EUR total compensation. The same behaviour is visible also for the 2007 event. Following the analysis stress testing A, B and C have been applied respectively focusing on the (i) evaluation of the data quality and of their accuracy to satisfy the Sendai Indicators calculation; (ii) evaluation of urban quality and resilience of the complex urban system of Benidorm when applied the 2007 event; and, (iii) applying the climate change impacts and urban growth projections on the urban system of Benidorm. Stress Testing A shows the lack of available fundamental data at the system level to support spatial and urban planning, considered as mitigation strategy to reduce exposure and vulnerability. Stress testing B reveals areas on which spatial planning should act to enhance the resilience of Benidorm to prepare, adsorb, recover and adapt different sectors (buildings, infrastructures, environment, touristic assets, and services) as well as their interdependencies. Stress testing C considers climate change projections developed within the municipal plans; it takes into account the commitment of the municipality in getting aligned with the challenges and needs posed by current and future risks. However, the investigation reveals that if no inversion of the current system dynamics would occur, a low performance would be experienced in adsorption, recovery phases when the system is impacted by stresses of climate change. Some suggestions can be drawn to support a proactive spatial planning by analysing those results as well.

5.1.1 Evaluation Framework

Table 5.1 provides an overview of the methodology and its application to the case-study. The evaluation is carried out by highlighting:

- the strengths of the applied method,
- opportunities for those who want to implement it,
- areas of improvement acknowledging some gaps of the presented research with the aim to devise strategies to overcome and enhance them for future works,
- as well as the related requirements.

5.2 Benidorm in the EU Risk Data Hub

The data of the Benidorm case study have been uploaded in the User corner data portal of the Risk Data Hub (RDH), with the agreement of the Spanish counterpart. The RDH platform developed in the Disaster Risk Management Knowledge Center (DRMKC) of the EU Commission JRC is a GIS web application which hosts, curates and disseminates data, tools and methodologies for Disaster Risk Management. It is composed by six main modules: the risk analysis data portal, the disaster loss data portal, a module hosting outcomes from various project developed in the field of disaster risk management, a module that delivers guidelines and instructions to navigate and make use of the platform, alongside explanatory reports about the methodologies implemented, the facts and figures section and the user corner module.

Once granted access rights, an external user can upload and manage its own disaster risk data both risk components (exposure, vulnerability and hazard) or/and disaster loss and damage data in the Country Corner. Loss data provided by the CCS have been rearranged and organized to be hosted by the platform according to the online provided guidance.[1]

The direct collaboration with platform's developers allowed to maintain the spatial reference characteristics of the data. In fact, the standard RDH structure allows to work, following the nomenclature of territorial units for statistics (NUTS) and local

---

[1] https://drmkc.jrc.ec.europa.eu/risk-data-hub#/external.

Table 5.1 Evaluation framework

	Methodology	Application
Strenghts	• Able to embed complexities of contemporary systems • Easily replicable and applicable to other case studies/contexts • Allows to consider the system in its whole (information, physical, economic, social domains and their interactions/interdependences) • Allows to consider different temporal scales (from pre-event to post-event considering the system behaviour in the prepare, adsorb, recover and adapt phases) • Applicable to system dimension (i.e. it does not limit the selection of a specific spatial scale of analysis) • Delivers results interpretable by experts in different disciplines and with different backgrounds	• Shows a real application of the stress testing • Combines heterogenous data available for the case study (maps, geospatial data, tabular data, text sections and narratives, economic values and so on) • Highlights local needs and criticalities to be addressed • Proposes suggestions to address a proactive spatial planning • Allows to monitor resilience in different time periods
Opportunities	• Allows to integrate and link specificities related to different disciplines (disaster and risks, climate change and spatial planning, data science) • Availability of an operational tool which allows a concrete evaluation of the resilience of urban systems • Allows to make the best use of current available data (qualitative and quantitative and of different formats) given its design • Simple but robust framework to be used also by non-specialists • Suitable for self-evaluation	• Results obtained can guide future choices and actions as well as local policy development • Benidorm Spatial planning can benefit from the revealed criticalities of the study • Local stakeholders can enhance their understanding and awareness • Represents a starting point for an open and constructive dialogue between play actors
Areas of improvement	• Being a novel approach, still not put in practice in a comprehensive way, it presents limitations that are not well known • Metrics to evaluate efficiency and to allow comparisons should be developed • Enhancement of the system performance measurement should be addressed developing scales which can be either tailored to the specific case study or constructed to be later compared for different contexts	• Further applications to different case studies would allow to properly identify areas of improvement • The development of the metrics and the score assignment could be addressed in cooperation with local stakeholders and organizations
Requirements	• Requires a large amount of data • Requires a good knowledge of the system under analysis • Metrics should be defined taking into account the necessity to have a multi-sectoral assessment • Requires a constant updating of the input data and of the metrics and their scores	

administrative units (LAU), on the following main scales: national (country), regional (basic regions-NUTS2), provincial (small regions-NUTS3) and municipal (LAU). Nevertheless considering that the loss data analysed are referred to the Postal Code level, a new geographical layer has been uploaded for Benidorm city so that a more refined spatial resolution could be used applying the main analysis functionalities provided by the platform. In addition data for the whole Province of Alicante have been uploaded aggregated at the municipality scale.

The region of analysis created on the platform refers to the Province of Alicante, from which it is possible to zoom and focus on Benidorm, where the data are specifically referred to the postal codes. For the area of analysis the "Assets" have been created, which refer to the assets that have been damaged and reported provided in the CCS dataset. Three type of metrics have been created:

- Losses adjusted with the GDP deflator, year of reference 2015 (Adjusted_losses_GDP_2015): this metric refers to the values of the losses adjusted with the GDP deflator, year of reference 2015. Through this metric it is possible to visualize data provided by the CCS adjusted through the GDP deflator method with 2015 as year of reference, and, the relative values in function of the GDP%00 computed according to the RDH methodology[2];
- Losses adjusted with the Consumer Price Index, year of reference 2021 (Adjusted_lossess_CPI_2021): this metric refers to the values of the losses adjusted with the Consumer Price Index (CPI) reference year 2021 provided by Spanish National Institute of Statistics.

The RDH system computes the data that are plotted on the interactive map according to the GDP of the province (source of the data EUROSTAT), adjusting first the data to the year of reference through the GDP deflator and then relating them to the specific GDP following the methodology developed within the RDH project.

Finally, as last step, data have been organized in excel file according to the template provided and uploaded through the administrative portal of the User Corner.

Main information regarding the uploading od the data are reported in the Table 5.2.

During the data upload, the platform generates a unique event code. Considering that floods may last more than a day and the definition of the event is not straightforward, the system collects the records with a temporal buffer of seven days.

In addition, for each included asset a Sendai Indicator code has been associated, more specifically those related to Target C[3] monitoring the direct economic loss provoked by disasters with respect to the global gross domestic product. An example of the output in the corner is shown in Fig. 5.1.

The Benidorm data were used to evaluate the Sendai indicators complementing the first stress test (Sect. 4.3). The stress test in fact evaluates the consistency and the comprehensiveness of the available data to measure the Sendai indicators thus permitting to evaluate the progress towards the targets set for disaster risk reduction.

---

[2] https://drmkc.jrc.ec.europa.eu/risk-data-hub#/methodologies.

[3] https://www.preventionweb.net/sendai-framework/sendai-framework-indicators.

Table 5.2 Main information of the of the dataset available in the user corner portal

Event_Code	Automatically generated by the RDH during the uploading of the data
Adm_Code	Administrative division: LAU (Province of Alicante); POSTAL CODES (Benidorm)
Year	Year the event took place (source CCS)
Begin_Date	Begin date of the event (date reported by CCS for each economic loss)
End_Date	End date of the event: equal to begin date of the event
Value_ Event	Numeric value of the losses (compensation provided by CCS in two formats according to the two metrics)
Hazard	Flood
Asset	Tractors and Agricultural Machinery; Automotive Industries; Big Surfaces; Commercial Activities; Construction Industries; Electrical and Electronic Industries; Food Industries; Hotel Facilities; Industrial Vehicles; Motorcycles; Offices; Other Assets; Other Industries; Residential Buildings; Sport Facilities; Textile Industries; Transport Infrastructures; Vehicles; Warehouses
Metric	Losses adjusted with the GDP deflator, year of reference2015 and Losses adjusted with the Consumer Price Index, year of reference 2021
Currency	Local
Data_ Source	CCS

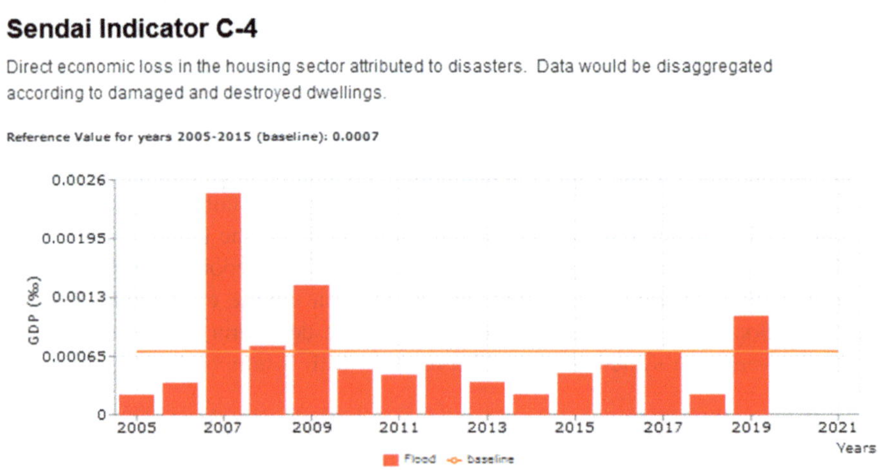

Fig. 5.1 An example of the final output of the Sendai Indicator Loss accounting

The definitive output is an interactive map that allows to filter the type of information to be plotted and returns infographics. Through the dedicated space an interactive map displays analysis at different spatial- and temporal-scales in addition to trends and measures of the Sendai Target C as in Fig. 5.2 where first it is shown how to select the data to plot and analyse and following all the assets data visualization first for the Alicante province and later for the municipality of Benidorm.

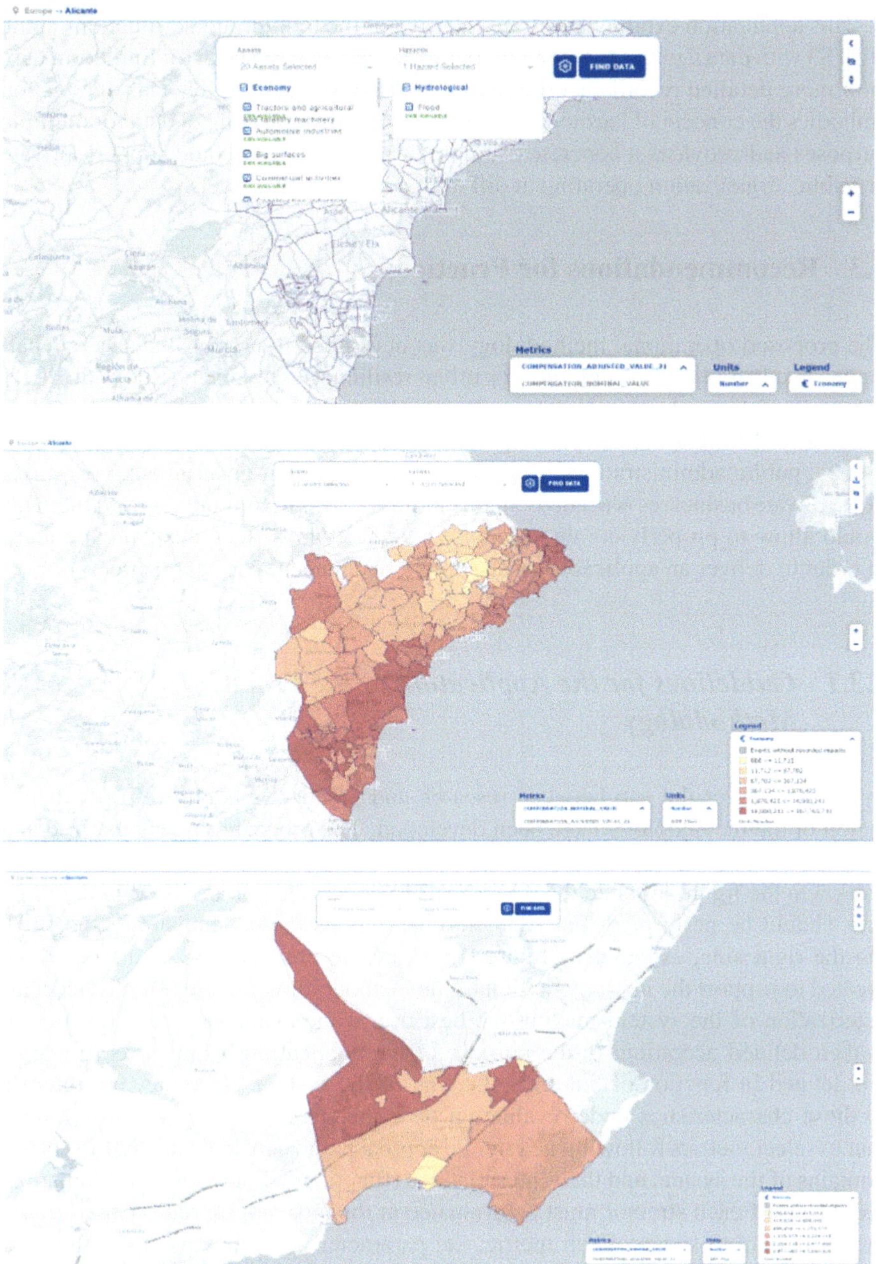

Fig. 5.2 The Alicante Province user corner interface

This application exhibits the "transcalarity" of the tool through evaluations at the NUTS3 with data aggregated at municipal scale; and, at LAU level for Benidorm City with more detailed records georeferenced at the postal code level. This application embodies the concept of harmonising, sharing, using and reusing of data for multiple purposes and provides a concrete example of how the RDH could support, private or public, organisation operating at different administrative levels.

5.3 Recommendations for Practice

The proposed operational methodology was developed to address the several challenges uncovered by the literature for urban resilience assessment (Rus et al. 2018; Ribeiro and Gonçalves 2019). In order to make the best use of the tool and to adapt it to different cities, the cooperation and collaboration of scientists, practitioners, public administrations, government agencies, non-governmental organizations, private businesses is needed (Shamsuddin 2020) as highlighted in the table. This would allow to properly combine scientific knowledge with contextual experience in order to deliver an applicable result linking resilience theory to practice.

5.3.1 Guidelines for the Application of the Proposed Methodology

In order to cover the gap between research and practice, guidelines for the application of the methodology have been developed. The procedure is schematized and illustrated in Fig. 5.3 so that it can be applied flexibly and pragmatically in future work. On the figure left side the seven steps to be carried out are listed in the order they should be performed. For each step the actions to accomplish are specified. On the right side, explanatory figures illustrate the practical use of the excel file created to support the implementation of the methodology. The first step is the characterization of the system making the best use of the available data. The stressor is then defined according to the purpose of the application, while several metrics are defined in function of the understanding of the system, since they are referred to those characteristics under evaluation once the stressor is applied. It is important to select metrics following a systemic approach that considers the multi-sectors/domains of the system and their interrelations (functions). Following the simulation, the effects of each stressor must be evaluated at the most appropriate scale for each metric assigning scores. Each metric, i.e. parameter considered, is then allocated in one of the cells of the resilience matrix and the scores assigned need to be then averaged in the final matrix (the average operation can be delineated according to the application, for instance it could be assigned an importance factor to each metric and then a weighted average can be performed). Once the resilience matrix is filled the system is assessed by evaluating its performance after the disruption has been applied.

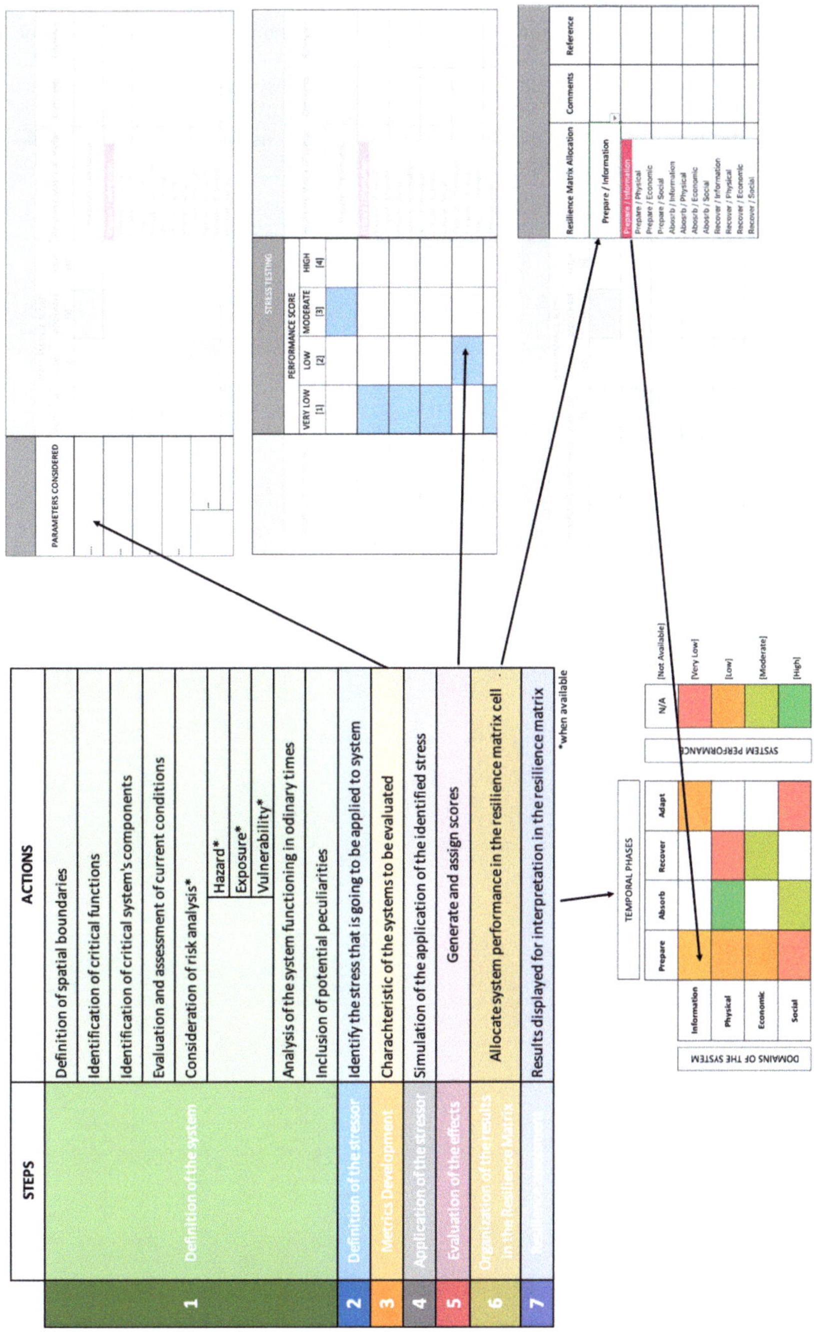

Fig. 5.3 Schematic guidelines for the application of the developed methodology

References

P.J.G. Ribeiro, L.A.P.J. Gonçalves, Urban resilience: a conceptual framework. Sustain. Cities Soc. **50**, 101625 (2019)

K. Rus, V. Kilar, D. Koren, Resilience assessment of complex urban systems to natural disasters: a new literature review. Int. J. Disast. Risk Reduct. **31**, 311–330 (2018)

S. Shamsuddin, Resilience resistance: the challenges and implications of urban resilience implementation. Cities **103**, 102763 (2020)

Chapter 6
Conclusions

This study aimed at developing an innovative practical and operational methodology, which considers damage and loss data analysis and uses the resilience matrix to organize stress testing results to answer questions as "How resilient is the urban system? What can be done to improve the planning system to secure more resilient urban systems?". The proposed methodology delivers an operational tool, which allows to consider the complexities of the urban area in its whole with its functions and across different temporal phases to understand how the system behaves when a stressor is applied and to identify critical areas where to allocate resources to enhance resilience. The methodology makes a step forward, it can take into account hazard, exposure and vulnerability but most importantly it allows to contemplate resilience. The latter is considered as a learning process through which the urban system evaluates the causes of past event damages. Future developments are hard to predict and the impact of disruptions are not always preventable, therefore urban systems should learn how to cope and live with known and unknown risks.

This work gets shaped around three pillars: damage and loss data, spatial planning and resilience. The proposed method embeds traditional risk assessments but goes beyond, addressing also recovery capacity, dynamic adaptation of cities building on knowledge fed by information and data that are systematically collected and updated each time a new event occurs and/or new lessons are learnt on how to prevent and manage extremes. Building resilience is a process which requires a strategic approach that aims at minimizing the negative impacts of adverse events on complex systems, improving their functionality by adapting to and learning from fundamental changes implied by the stress faced. The methodology highlights the connection between the past and the future showing how the former influences the future and how knowledge of the past can improve configurations of the future. Th proposed method is a first attempt to tackle the growing challenges deriving from climate change and the limited knowledge that we still have regarding its cross-scale impacts, given the increased interconnectedness of economies and urban in multi-hazard and multi-risk conditions.

A. Faiella, *Learning from Disasters to Enhance Resilience of Urban Complex Systems*, PoliMI SpringerBriefs, https://doi.org/10.1007/978-3-032-07356-3_6

Resilience approaches become an effective solution for territorial systems, being able to entail dynamicity and flexibility, recognizing system complexities while focusing on how these dynamics and feedbacks evolve across temporal and spatial scales learning from both successes and failures. Such approach that envisages a cross- and multi-domain analysis is challenging due to the required volume of data related to a variety of different fields for a proper implementation. Despite recent policies and rapid technological advancement "Having no data is all too often the case, whether because no relevant data exist; they exist but cannot be found; exist but are not available due to proprietary control, embargoes, technical barriers, degradation due to lack of curation; or simply because those who have the data cannot or will not share them" as pointed out by Borgman (2017). Thus, we still record a lack of basic figures of disaster losses at the scale, granularity and level of detail needed for most applications (Faiella et al. 2022).

Whilst the methodology requires a large amount of data, it does not require a specific data format, and permits to exploit heterogenous and assorted data as it is adaptable to different data formats: numerical, textual, satellite images, maps and so on. Such flexibility is possible thanks to the fact that each input can be converted in a metric intended to represent a characteristic of the system, which allows to have different data and information comparable; and, once the stressor is applied, it is assessed to answer the question "how good is the system performance according to that metric when a stressor x is applied?".

Additionally, the methodology can be implemented in different territorial systems, of any size, by a variety of users with different skills and backgrounds as for instance administrations, organizations, researchers, urban planners, citizens. Results, organized as a final output within the resilience matrix, are also of straight interpretation so that they can be shared and discussed with different stakeholder to guide further actions on those variables that should be improved as highlighted in the matrix and in the stress test. The process itself can guide in the selection of potential projects, mitigation measures, revision of spatial planning decisions towards resilience according to the results of the methodology application.

Being both the resilience matrix and stress testing a novel approach and pioneering tools, they still require some consolidation. Metrics to evaluate resilience management efficiency and to allow comparisons within different strategies are still in the early stages of development (Linkov et al. 2013; Henry and Ramirez-Marquez 2012; Jones 2019). Constant updating of the input data and of the metrics and their scores is also required to detect and monitor changes in the urban environment.

The knowledge that derives from the use of the methodology may better support spatial planning (Fig. 6.1), as it provides the set of information on ecosystems, agriculture, economy, soil consumption, mobility, demography, historical heritage and identity, necessary for an organic and exhaustive representation and evaluation of the state of the territory. The latter is in fact the compound of various functions and the result of evolutionary processes that characterize each city in a unique way with respect to objectives such as resilience, sustainability, and disaster risk reduction.

Specifically the application of the methodology to the case study and its procedure clearly revealed matters of Benidorm's urban pattern that should have been or should

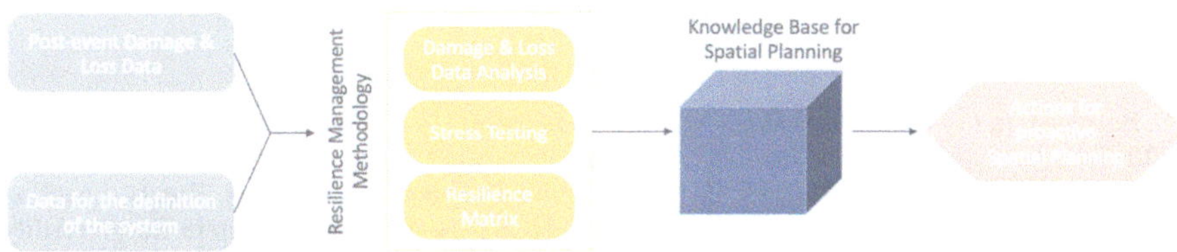

Fig. 6.1 Process of spatial planning knowledge enhancement through resilience management

be addressed to achieve more resilient configurations. Mainly those matters can be classifed in two principal clusters: (i) data scarcity, and (ii) absence of structured tools or procedures to support capitalization of knowledge needed to foster proactive spatial planning.

The present research immediately faced scarcity of data. Official municipal documents introduce the presence of different hazards in the area of Benidorm but they do not handle or address them any further. The only hazard for which a more consistent volume of data has been found is flooding, thus as a result of the PATRICOVA which followed the EU Flood Directives (2007/60/EC). Vulnerability is never contemplated and exposure is inadequately addressed. Therefore, it becomes clear that any seen anthropic development ignored the real risks in the area.

Absence of tools or procedures is the result of both scarcity of data but also of inability to learn from past experiences. The case study clearly shows a system which fails to learn from past events; no lessons from experienced events have been drawn. If damage and loss data available were properly systematized they could have conveyed enhanced understanding of the city's criticalities encouraging the implementation of mitigation measures aimed at reducing flood risk. Instead over the years the same damages and losses occurred in the same areas affected by previous floods. Besides the development of areas which were not considered developable in the first Master Plan shows the absence of a vision that considers strategically the hazardous characteristics of the city. The lack of comprehensive spatial planning which is able to envision longer-term strategies for the municipality is an important issue to be addressed especially when considering the foreseen impacts of climate change on the area at stake.

The application of the proposed methodology provides a strategic approach to be formally embedded into planning approaches and institutional frameworks. As shown in the case of Benidorm, making proper use of knowledge from past events through a defined and structured methodology may contribute to bridge the gap between real territorial needs and planning to enable the latter, as expressed by Fridemann (1998, p.250), to "continuously reinvent itself" as circumstances—now more than ever constantly—change.

References

C.L. Borgman, in *Big Data, Little Data, No Data: Scholarship in the Networked World* (MIT press, 2017)

A. Faiella, S. Menoni, M.P. Boni, M. Panoutsopoulou, T. Thoma, S. Salari, N. Rueda, Enabling knowledge through structured disaster damage and loss data management system. Sustainability **14**(10), 6187 (2022)

J. Friedmann, Planning theory revisited. Eur. Plan. Stud. **6**(3), 245–253 (1998)

D. Henry, J.E. Ramirez-Marquez, Generic metrics and quantitative approaches for system resilience as a function of time. Reliab. Eng. Syst. Saf. **99**, 114–122 (2012)

L. Jones, Resilience isn't the same for all: Comparing subjective and objective approaches to resilience measurement. Wiley Interdisciplinary Reviews: Clim. Change **10**(1), e552 (2019)

I. Linkov, D.A. Eisenberg, K. Plourde, T.P. Seager, J. Allen, A. Kott, Resilience metrics for cyber systems. Environ. Syst. Decis. **33**(4), 471–476 (2013)

MIX
Papier aus verantwortungsvollen Quellen
Paper from responsible sources
FSC® C105338

If you have any concerns about our products,
you can contact us on
ProductSafety@springernature.com

In case Publisher is established outside the EU,
the EU authorized representative is:
Springer Nature Customer Service Center GmbH
Europaplatz 3, 69115 Heidelberg, Germany

Printed by Libri Plureos GmbH
in Hamburg, Germany